Investigating Matter Through Inquiry

A project of the American Chemical Society
Education and International Activities Division
Office of K–8 Science

American Chemical Society

American Chemical Society

Inquiry in Action—Investigating Matter Through Inquiry

The activities described in this book are intended for students under the direct supervision of teachers. The American Chemical Society cannot be held responsible for any accidents or injuries that might result from conducting the activities without proper supervision, from not specifically following directions, or from ignoring the cautions contained in the text.

American Chemical Society Staff

Authors
James H. Kessler
Patricia M. Galvan

Education and International Activities Division Director, Sylvia A. Ware
Associate Director for Academic Programs, Michael J. Tinnesand
Cover, Neal R. Clodfelter
Designer, Peggy Corrigan
Copy Editor, Kelley Carpenter
Illustrations, Jim Wisniewski

Safety Review—ACS Committee on Chemical Safety

Library of Congress Cataloging-in-Publication Data

Kessler, James H.
Inquiry in action : investigating matter through inquiry / James H. Kessler, Patricia M. Galvan.
p. cm.
"A project of the American Chemical Society, Education and International Activities Division, Office of K-8 Science."
ISBN 0-8412-3871-5
1. Science—Study and teaching (Elementary)—Activity programs. 2. Science—Study and teaching (Middle school)—Activity programs. 3. Matter—Study and teaching (Elementary)—Activity programs. 4. Matter—Study and teaching (Middle school)—Activity programs. 5. Matter—Experiments. I. Galvan, Patricia M., 1967 II. American Chemical Society. Education and International Activities Division. Office of K–8 Science. III. Title.

LB1585.K474 2003
372.3'5—dc21

2003043704

Investigating Matter Through Inquiry

For information about ACS Education and
International Activities Division products and programs,
visit chemistry.org, click on Educators and Students, and scroll to K–12.

Table of contents

Introduction

Welcome to *Inquiry in Action—Investigating Matter Through Inquiry*. The purpose of *Inquiry in Action* is to give elementary and middle-school teachers a set of physical science activities to help teach the major concepts in the study of matter. The activities were developed to lend themselves to a guided-inquiry approach and to work across the range of grades 3–8. To be effective over such a wide grade range, the activities are designed to cover basic concepts but have the flexibility to be modified by teachers through varying questioning strategies, the degree of guidance given students, and the vocabulary used. The materials for all activities are very common, safe, inexpensive, and available at any grocery store.

To help determine what concepts to include within the topic of matter, we based our decisions on the *National Science Education Content Standards for Physical Science and Science as Inquiry* and also on the results of teacher surveys asking for the content most often studied within the topic of matter. Based on the Standards and teacher input, the physical science topics covered include Scientific Questions and Their Investigation, Physical Properties, Physical Change, Chemical Change, States of Matter, Density, and Mixtures and Solutions.

The presentation method used in *Inquiry in Action* is a guided-inquiry approach. Each investigation begins with a teacher demonstration or student observation that serves as motivation for either student or teacher questioning. Together, the teacher and students develop a question to investigate and begin to design an experiment to answer the question. In most investigations, this initial designing and conducting of experiments is done with substantial teacher guidance. This allows teachers to model the thinking processes involved in investigating a scientific question and gives students some familiarity with the science concepts and a context for further investigation. As students gain experience, they take on more responsibility in designing experiments later in the investigation to answer related questions.

The activities in *Inquiry in Action* include many suggestions for questioning strategies before, during, and after the activities. Each activity also includes examples of experimental procedures with all the required materials, expected results, and assessment ideas. All these suggestions and examples can serve as a guide as teachers develop the investigations with students. The spontaneous nature of inquiry, individual teaching styles, and the variety of potential responses from students will determine how each activity and investigation is actually conducted.

Field testing *Inquiry in Action* activities with students and at numerous teacher workshops helped tremendously in refining the activities themselves as well as the guided-inquiry approach to presenting them. Also, many useful suggestions were incorporated from very thorough and thoughtful reviews supplied by National Science Teachers Association (NSTA) reviewers and editorial staff.

Our hope is that the guided-inquiry approach in *Inquiry in Action* will help teachers think about trying new or traditional activities in a way that will motivate students to explore and achieve the physical science and inquiry goals of today's science standards.

The *National Science Education Standards*

Inquiry

Many science programs place a high priority on students doing hands-on science. However, there are times when students will participate fully in science activities yet have little understanding of the concepts covered or of the scientific processes used. The *National Science Education Standards*, National Research Council, Washington, DC: National Academy Press, 1996, address these problems by focusing on the importance of *inquiry*.

In the *Standards*, the term *inquiry* is used in two different ways. It describes both what students need to learn and how teachers need to teach. As applied to students, the content standard of "science as inquiry" comprises a set of abilities and understandings that students develop as they meaningfully participate in the processes of scientific investigation. "Science as inquiry," in this sense, deals with student questions, observation, measurement, experimental design, logical reasoning based on evidence, and communicating results.

As applied to teachers, inquiry or inquiry-based teaching refers to the strategies and techniques that teachers use to engage and guide students through scientific investigations. Teachers should involve students as much as possible in the entire process of conducting a scientific investigation. In fact, whenever possible, students' own questions and interests should initiate an investigation. One of the most notable features of inquiry is that teachers help students identify and communicate the thought processes involved in designing and conducting the investigation. Through this experience with inquiry-based teaching and learning, students can acquire both the skills of scientific inquiry and the concepts of the physical science curriculum.

Inquiry can have many variations depending on the needs, experience, and prior knowledge of students, the time and resources available, and the concepts being taught. The five essential features of inquiry on page 2 and the chart on page 3 are adapted from the book *Inquiry and the National Science Education Standards, A Guide for Teaching and Learning*. National Research Council, Washington, DC: National Academy Press, 2000. The five features and chart show the variety of approaches to teacher and student involvement and input in inquiry-based teaching and learning. More student-centered or "open" inquiry is to the left and more teacher-centered, or "guided" inquiry is to the right.

Concerning the more "open" inquiry, *Inquiry and the National Science Education Standards* states, *"...students rarely have the ability to begin here. They first have to learn to ask and evaluate questions that can be investigated, what the difference is between evidence and opinion, how to develop a defensible explanation, and so on. A more structured type of teaching develops students' abilities to inquire.... Experiences that vary in "openness" are needed to develop inquiry abilities. Students should have opportunities to participate in all types of inquiries in the course of their science learning"* (pages 29–30).

Inquiry in Action encourages teachers to view inquiry in this same way—as a flexible approach that can be incorporated into science teaching in a variety of ways to improve student learning.

The full text of the *National Science Education Standards* and *Inquiry and the National Science Education Standards* can be found at the National Academy Press at http://www.nap.edu.

The following information and charts on pages 2–8 are adapted from *Inquiry and the National Science Education Standards* (pages 18–30) and the *National Science Education Standards* (pages 127 and 154).

Essential features of inquiry

Inquiry-based learning has five essential features that apply across all grade levels. Although not all features must be present in each lesson, each should be present at some time over the course of a series of lessons.

1. Students begin with a question that can be answered in a scientific way.

Questions that can be investigated in an elementary or middle-school classroom develop in a variety of ways. Most commonly, teachers provide opportunities that invite student questions by demonstrating a phenomenon or having students engage in an open investigation of objects, substances, or processes. Sometimes, questions will develop from something the students observe and suggest. Other times, the teacher provides the question. Either way, questions must be able to be investigated in a developmentally appropriate way in an elementary or middle-school classroom. Teachers will likely have to modify student questions into ones that can be answered by students with the resources available, while being mindful of the curriculum.

2. Students rely on evidence in attempting to answer the question.

This evidence can come from designing and conducting an investigation; observing a teacher demonstration; collecting specimens; or observing and describing objects, organisms, or events. The evidence can also come from books or electronic media.

3. Students form an explanation to answer the question based on the evidence collected.

Scientific explanations provide causes for effects and establish relationships based on evidence and logical argument. For students, scientific explanations go beyond current knowledge to build new ideas upon their current understanding.

4. Students evaluate their explanation.

Students consider questions such as the following: Can other reasonable explanations be based on the same evidence? Are there any flaws in the reasoning connecting the evidence to the explanation?

5. Students communicate and justify their proposed explanations.

Sharing explanations can help strengthen or bring into question students' procedures as well as their reasoning in connecting the evidence from their experiments to their explanations.

These five features of scientific inquiry do not need to happen in a formal and strict step-by-step sequence. Some features may cycle a few times as students learn about the topic or concept and generate new questions. These elements merely guide the process of doing science and should unfold during the course of an activity or series of activities and discussions. If all of the features are present in an activity, the lesson is considered to be full inquiry. Partial inquiry is acceptable as long as students gain experience with all of the features at some time. The role of the teacher and of the students can shift, as well. At times, the investigation is more student-directed, or open, and at times the teacher takes a more direct role in guiding the lessons. All of these variations are valid inquiry.

Inquiry and its variations

Student role	Variations			
engages in scientifically oriented *questions*	poses a question	selects among questions, poses new questions	sharpens or clarifies question provided by teacher, materials, or other source	engages in question provided by teacher, materials, or other source
gives priority to evidence in responding to questions	determines what constitutes evidence and collects it	directed to collect certain data	given data and asked to analyze	given data and told how to analyze
formulates explanations from evidence	formulates explanation after summarizing evidence	guided in process of formulating explanations from evidence	given possible ways to use evidence to formulate explanation	provided with evidence
connects explanations to scientific knowledge	independently examines other resources and forms the link to explanations	directed toward areas and sources of scientific knowledge	given possible connections	
communicates and justifies explanations	forms reasonable and logical argument to communicate explanations	coached in development of communication	provided broad guidelines to sharpen communication	given steps and procedures for communication
	More ◀—— Amount of Learner Self-Direction ——▶ Less Less ◀—— Amount of Direction from Teacher or Material ——▶ More			

Selected Science Teaching Standards

Teaching Standard A
Teachers of science plan an inquiry-based science program for their students.

Develop a framework of yearlong and short-term goals for students.

- Teachers adapt school and district program goals, as well as state and national goals, to the experiences and interests of their students individually and as a group.
- A challenge to teachers of science is to balance and integrate immediate needs with the intentions of the yearlong framework of goals.
- The content standards, as well as state, district, and school frameworks, provide guides for teachers as they select specific science topics.
- In planning and choosing curricula, teachers strive to balance breadth of topics with depth of understanding.

Select science content and adapt and design curricula to meet the interests, knowledge, understanding, abilities, and experiences of students.

- Whether working with mandated content and activities, selecting from extant activities, or creating original activities, teachers plan to meet the particular interests, knowledge, and skills of their students and build on their questions and ideas.
- Teachers are aware of and understand common naïve concepts in science for given grade levels, as well as the cultural and experiential background of students and the effects these have on learning.

Select teaching and assessment strategies that support the development of student understanding and nurture a community of learners.

- Inquiry into authentic questions generated from student experiences is the central strategy for teaching science.
- Teachers focus inquiry predominantly on real phenomena, in classrooms, outdoors, or in laboratory settings, where students are given investigations or guided toward fashioning investigations that are demanding but within their capabilities.

Teaching Standard B
Teachers of science guide and facilitate learning.

Focus and support inquiries while interacting with students.

- Teachers and students collaborate in the pursuit of ideas, and students quite often initiate new activities related to an inquiry.
- Teachers match their actions to the particular needs of students, deciding when and how to guide—when to demand more rigorous grappling by the students, when to provide information, when to provide particular tools, and when to connect students to other sources.
- Teachers continually create opportunities that challenge students and promote inquiry by asking questions.

- Although open exploration is useful for students when they encounter new materials and phenomena, teachers need to intervene to focus and challenge the students, or the exploration might not lead to understanding.
- Teachers must decide when to challenge students to make sense of their experiences and ask students to explain, clarify, and critically examine and assess their work.

Orchestrate discourse among students about scientific ideas.

- Teachers encourage oral and written discourse that focuses the attention of students on how they know what they know and how their knowledge connects to larger ideas, other domains, and the world beyond the classroom.
- Teachers require students to record their work—teaching the necessary skills as appropriate—and promote many different forms of communication (for example, spoken, written, pictorial, graphic, mathematical, and electronic).
- Teachers assist students to work together in small groups and give groups opportunities to make presentations of their work to explain, clarify, and justify what they have learned.

Challenge students to accept and share responsibility for their own learning.

- Teachers give individual students active roles in the design and implementation of investigations, in the preparation and presentation of student work to their peers, and in student assessment of their own work.

Encourage and model the skills of scientific inquiry, as well as the curiosity, openness to new ideas, and skepticism that characterize science.

- A teacher who engages in inquiry with students models the skills needed for inquiry.
- Teachers who exhibit enthusiasm and interest and who speak to the power and beauty of scientific understanding instill in their students some of those same attitudes toward science.

Teaching Standard C
Teachers of science engage in ongoing assessment of their teaching and of student learning.

Use multiple methods and systematically gather data on student understanding and ability.

- Teachers carefully select and use assessment tasks that are also good learning experiences.
- Teachers observe and listen to students as they work individually and in groups. They interview students and require formal performance tasks, investigative reports, written reports, pictorial work, models, inventions, and other creative expressions of understanding.
- Teachers examine portfolios of student work, as well as more traditional paper-and-pencil tests.

For frequently asked questions about inquiry-based teaching and learning, see *Frequently Asked Questions About Inquiry*, page 190.

Inquiry and physical science content standards

Science as inquiry: Abilities necessary to do scientific inquiry

The *Standards* describe a set of abilities that enables students to meaningfully participate in the processes of scientific investigation. These abilities of scientific inquiry focus on student questions, observation, measurement, experimental design, logical reasoning based on evidence, and communicating results.

K–4	5–8
Ask a question.	Identify questions that can be answered through scientific investigations.
Plan and conduct a simple investigation.	Design and conduct a scientific investigation.
Employ simple equipment and tools to gather data and extend the senses.	Use mathematics in all aspects of scientific inquiry. Use appropriate tools and techniques to gather, analyze, and interpret data.
Use data to construct a reasonable explanation.	Develop descriptions, explanations, predictions, and models using evidence.
	Think critically and logically to make the relationships between evidence and explanation.
	Recognize and analyze alternative explanations and predictions.
Communicate investigations and explanations.	Communicate scientific procedures and explanations.

Science as inquiry: Understandings about scientific inquiry

As students become active participants in scientific investigations, they will develop certain ideas about science. They will realize that certain kinds of questions can be answered in a scientific way. They will also come to understand that scientific investigations use a logical approach to answer a question. The cumulative experience of inquiry-based teaching and learning will give students a fuller understanding of the work of scientists.

<table>
<tr><th>K–4</th><th>5–8</th></tr>
<tr><td>Scientific investigations involve asking and answering a question.</td><td>Different kinds of questions suggest different kinds of scientific investigations.</td></tr>
<tr><td>Scientists use different kinds of investigations depending on the questions they are trying to answer.</td><td>Current scientific knowledge and understanding guide scientific investigations.</td></tr>
<tr><td rowspan="2">Simple instruments, such as magnifiers, thermometers, and rulers, provide more information than scientists obtain using only their senses.</td><td>Mathematics is important in all aspects of scientific inquiry.</td></tr>
<tr><td>Technology used to gather data enhances accuracy.</td></tr>
<tr><td rowspan="2">Scientists develop explanations using observations (evidence) and what they already know about the world (scientific knowledge).</td><td>Scientific explanations emphasize evidence and have logically consistent arguments.</td></tr>
<tr><td>Science advances through legitimate skepticism.</td></tr>
<tr><td>Scientists make the results of their investigations public.</td><td>Scientific investigations sometimes result in new ideas and phenomena for study.</td></tr>
<tr><td>Scientists review and ask questions about the results of other scientists' work.</td><td>Scientists review and ask questions about the results of other scientists' work.</td></tr>
</table>

Physical science: Properties of objects and materials Properties and changes of properties in matter

The *Standards* place most topics, traditionally considered chemistry-related, under the physical science content standard of properties of objects and materials for grades K–4 and properties and changes of properties in matter for grades 5–8. The topics addressed under these content standards are summarized in the chart below.

K–4	5–8
Objects have observable properties, including size, weight, and temperature, which can be measured using tools such as rulers, balances, and thermometers.	A substance has characteristic properties such as density, a boiling point, and solubility, which are all independent of the amount of the sample.
	Substances react chemically in characteristic ways with other substances to form new substances with different characteristic properties.
Objects can be described by the properties of the materials from which they are made and can be separated or sorted according to those properties.	A mixture of substances can often be separated into the original substances using one or more of the characteristic properties.
	Substances are often placed in categories or groups if they react in similar ways.
Materials can exist in different states—solid, liquid, and gas—and can be changed from one state to another by heating and cooling.	
	There are more than 100 known elements that combine in a multitude of ways to produce compounds, which account for the living and nonliving things we encounter.
	In chemical reactions, the total mass is conserved.

The physical science topics in Inquiry in Action

Most of the concepts in the chart on the left are addressed under the following topics covered in *Inquiry in Action*: **physical properties**, **physical change**, **chemical change**, **states of matter**, **density**, and **mixtures and solutions**.

To facilitate student understanding of the concepts in the chart, the standards strongly encourage student participation in the design of investigations. As students design and conduct investigations into objects, substances, and phenomena, they will observe and measure various properties and processes of matter. The degree of sophistication with which students can conduct these investigations will depend largely on the grade level of the students and their experiences doing science activities.

Many, if not all, of the properties of objects, materials, and substances listed in the chart can be explained by referring to the atoms and molecules of which they are made. The type of atoms and molecules that make up a substance and the arrangement and strength of the bonds between them determine the different characteristic properties of that substance including its physical properties and how it undergoes physical and chemical change. But for teachers at the 3–8 grade level, for whom *Inquiry in Action* is written, it is not necessary or advisable to focus on the atomic and molecular explanation of *why* substances have the characteristic properties they do. It is better for students in this grade range to concentrate on their own observations of the different characteristics of substances by investigating them directly.

The *National Science Education Standards* (page 149) state: *It can be tempting to introduce atoms and molecules or improve students' understanding of them so that particles can be used as an explanation for the properties of elements and compounds. However, use of such terminology is premature for these students and can distract from the understanding that can be gained from focusing on the observation and description of macroscopic features of substances and of physical and chemical reactions.*

Features of *Inquiry in Action* investigations and activities

Safety

All activities in *Inquiry in Action* have been reviewed for safety. Although only household substances and equipment are used, all participants should wear safety goggles when doing any of the activities. For more information on safety, call the American Chemical Society at 1-800-227-5558 for the free booklet *Safety in the Elementary Science Classroom.*

Materials chart

The materials for each activity in an investigation are listed in one chart at the beginning of the investigation. The materials included in this chart are those needed if the suggested procedures written for each activity are conducted exactly as written. Depending on how you present the lesson and students' ideas, you may need other materials. The materials charts are intended to give you a good idea of the materials you will need for one group to conduct each activity within the investigation. When gathering the materials for each activity, multiply the amounts listed by the number of groups that you will have working in your class.

Teacher preparation

If any of the activities in an investigation require advance preparation, instructions are given before the first activity in the investigation.

Science background information

Each investigation includes science background information about the topic being investigated. Explanations on the molecular level are also included for each activity so that teachers will have a quick reference for better understanding the phenomena observed in the activities.

Activities

Each investigation consists of a series of activities that build upon one another to fully develop the science concepts. In the beginning of each investigation, students either do an introductory activity or see a demonstration that shows an interesting phenomenon related to the topic being investigated. This serves as a way for teachers to assess students' prior knowledge and experience with the topic. The introductory activity also gives students a common context in which to ask questions and motivates them to investigate some of these questions in the activities that follow.

An important part of the activities are the suggested questions included for teachers to ask students. These questions are provided as samples that can be used to engage students in the activity. Questions will cover identifying and controlling variables, designing experiments, communicating observations, logically connecting evidence to explanations, and evaluating results. The questions are not written as a script but as suggestions for teachers to use as needed, along with other questions that arise naturally during the course of the activity.

Question to investigate

A question at the beginning of each activity serves as a guide for designing and conducting the experiments to answer the question. All the activities within an investigation lead up to or contribute to the overall objective of the investigation.

Procedure

Procedures are examples or models of ways to design and conduct an experiment to answer the question to investigate. Since student input in the experimental design is encouraged, the procedure may be modified or completely changed. The procedure is offered as a guide to show one way an activity can be conducted.

Expected results

The expected results are intended to provide an idea of likely outcomes for each activity.

Activity sheets

Each investigation includes activity sheets to record observations and to help focus and assess learning.

Assessment Rubric

At the end of each investigation is a scoring rubric to help evaluate student responses on activity sheets and for helping to assess student participation in the investigation. Teachers can circle items from this list and add others to give students more feedback on their work.

Investigation 1.
Scientific questions and their investigation

Summary

Students will observe what happens when an M&M is placed in a shallow plate of water. After making observations about the movement of the color in the water, students will begin to ask other related questions and investigate them. Student questions and suggestions will lead to investigations exploring the variables that affect the movement of the color. Student questions will probably involve variables such as the color of the M&Ms, number of M&Ms, where they are located in the plate, temperature of the water, and the type of solution the M&Ms are placed in.

Objective

Students will begin to recognize the characteristics of a scientific question. They will ask questions that they can investigate, make predictions, conduct tests, and record their observations. Students will use the term *properties* in reference to the characteristic way color comes off an M&M in water.

Assessment

The assessment rubric *Scientific questions and their investigation* on page 34 is included so that you can assess and document student progress throughout the investigation. The abilities and understandings demonstrated by students and recorded on the rubric include the following: Students should be able to recognize and begin to formulate questions that they can investigate in a scientific way, make a plan to investigate the question, record observations with simple drawings, and if possible answer the question based on the results of the experiment. Investigative behaviors observed as students plan and conduct their investigations, communicate their observations, and work with their groups are also recorded on the rubric.

Relevant *National Science Education Standards*

Physical Science

K–4

Properties of Objects and Materials

Objects have many observable properties.

5–8

Properties and Changes of Properties in Matter

A substance has characteristic properties.

Science as Inquiry

K–4

Abilities Necessary to do Scientific Inquiry

Ask a question about objects.
Plan and conduct a simple investigation.
Employ simple equipment and tools to gather and extend the senses.
Use data to construct a reasonable explanation.
Communicate investigations and explanations.

Understandings about Scientific Inquiry

Scientific investigations involve asking and answering a question.
Types of investigations include describing objects…and doing a fair test.
Good explanations are based on evidence from investigations.

5–8

Abilities Necessary to do Scientific Inquiry

Identify questions that can be answered through scientific investigations.
Design and conduct a scientific investigation.
Use appropriate tools and techniques to gather, analyze, and interpret data.
Develop descriptions, explanations, predictions, and models using evidence.
Think critically and logically to make the relationships between evidence and explanations.
Communicate scientific procedures and explanations.

Understandings about Scientific Inquiry

Different kinds of questions suggest different kinds of scientific investigations.
Scientific explanations emphasize evidence and have logically consistent arguments.
Scientific investigations sometimes result in new ideas and phenomena for study or generate new procedures for an investigation. These can lead to new investigations.

How this investigation relates to the *Standards*

The *Standards* state that students should be able to ask a question that can be answered in a scientific way. It is very easy for students to ask questions, but it is often difficult for them to ask a question that can be answered by a scientific investigation. It is even harder for them to ask a question that *they* can answer with the resources available to them. With teacher guidance, students will be able to refine their questions and design scientific investigations to attempt to answer them. Through this experience, students will begin to recognize the types of questions that can be answered in a scientific way and will begin to develop an understanding of variables and the need to control them to design a fair test.

Materials list

1.1	One M&M in water
1.2	Racing M&M colors
1.3a	Two M&Ms
1.3b	M&Ms in different arrangements
1.4.	M&M colors in different temperatures
1.5.	M&M colors in different liquids

Each group will need

	Activities					
	1.1	1. 2	1.3a	1.3b	1.4	1.5
M&Ms	•	•	•	•	•	•
white plastic or foam dinner plates	1	7	2	≥ 1	3	3
room-temperature water	•	•	•	•	•	•
hot tap water					•	
cold water					•	
3½-ounce plastic cups	1	1	1	1	1	3
9- or 10-ounce plastic cups						2
crayons or colored pencils	•	•	•	•	•	•
permanent marker		•			•	•
sugar						•
salt						•
plastic teaspoon						1
bucket or bowl	•	•	•	•	•	•
paper towels	•	•	•	•	•	•

Notes about the materials

Throughout the investigation, M&Ms are used but you can use any color-coated candy that gives similar results. It may be interesting to try others.

Students will use the bucket or bowl to empty the liquid from the plates after each investigation. They will then dry the plates with paper towels so that they can be used again. The crayons and colored pencils will be used with *Activity sheet 1.2, 1.3, 1.4, 1.5—Investigating questions about M&Ms in water* (page 33).

Teacher preparation

Give each group one small bag of M&Ms to use for all five activities. The following are the minimum number and type of M&Ms needed in each activity:

Activity 1.1	1 M&M
Activity 1.2	6 different colors
Activity 1.3a	2 of the same color and 1 of a different color
Activity 1.3b	mix of different colors number to be determined by the experiments students select
Activity 1.4	3 of the same color
Activity 1.5	3 of the same color

Activity sheets

Copy the following activity sheets for this investigation and distribute them as specified in the activities.

Activity sheet 1.1 — page 32
One M&M in water

Copy one per student.

On this activity sheet, students will record some of the questions they might be able to investigate. They will then select one question and make a preliminary plan to investigate the question.

Activity sheet 1.2, 1.3, 1.4, 1.5 — page 33
Investigating questions about M&Ms in water

Copy one per student, perhaps more if students conduct more than five M&M activities.

Students will use this activity sheet as they conduct each of the activities. They will write questions that can be investigated in a scientific way, record their observations with a drawing, and if possible answer the question they were investigating.

Science background information

The activities in Investigation 1 and many other activities in *Inquiry in Action* use *water* as a major ingredient. That's because many chemistry-related physical science phenomena are due to the unique properties of water molecules and their interaction with other molecules. It may have been a few years since your last chemistry class but if you spend a little time reading the following explanation, it may help you better understand the observations that you and your students will make during the activities in Investigation 1 and many of the activities in later investigations. These explanations are intended for your own background and not as a basis for explaining the activities and observations to elementary or middle-school students. Explanations of physical and chemical change on the molecular level can be very challenging for elementary and middle-school students to understand.

The molecular structure of water molecules

A water molecule, or H_2O, is made up of two *hydrogen atoms bonded* to one *oxygen atom*. It is the special character of this oxygen–hydrogen bond that gives water many of its unique properties.

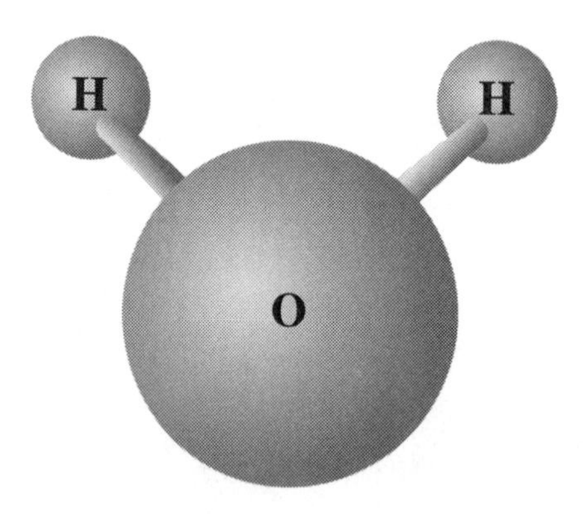

Atoms can form a covalent bond

Atoms have a certain number of protons in their nucleus and a corresponding number of electrons around the nucleus. Because the nucleus contains protons, which have a positive charge, the nucleus has an attraction for electrons, which have a negative charge. When two atoms get near each other, the nucleus of each atom has a certain amount of attraction for the electrons of the other atom. If the attractions are just right, one or more electrons from each atom end up being attracted by both atoms instead of just the atom the electrons originally started with. When this happens, it is called a *covalent bond* between the atoms. A covalent bond is sometimes referred to as a "sharing" of electrons between the bonded atoms.

Oxygen and hydrogen form a covalent bond

Because of its atomic make-up, oxygen has a strong attraction for electrons. Hydrogen can also attract electrons but not as strongly as oxygen does. When oxygen and hydrogen get near each other, they can share a mutual attraction for electrons and form a covalent bond. The gray "cloud" around the oxygen and hydrogen atoms represents the area where electrons are likely to be found in the covalently bonded water molecule. This covalent bond between oxygen and hydrogen in the water molecule gives water many of its special characteristics.

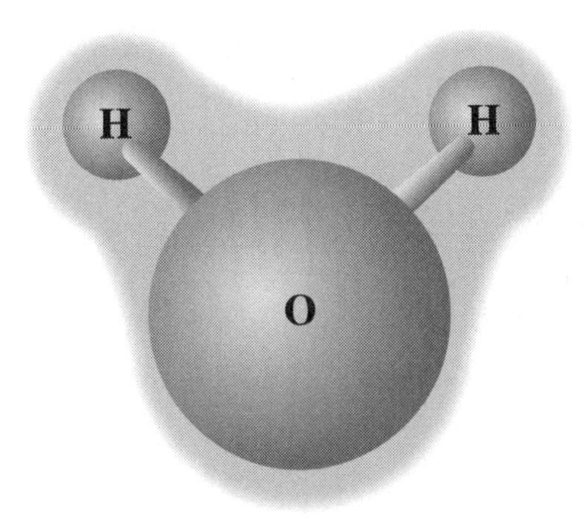

The polar nature of the oxygen–hydrogen bond

Here comes the part that makes water so special. Because the oxygen has a greater attraction for electrons than the hydrogens have, the electrons shared between the oxygen and hydrogen atoms spend more time around the oxygen than they do around the hydrogens. Since electrons have a negative charge, this makes the area around the oxygen atom slightly negatively charged and the area around the hydrogen atoms slightly positively charged. When a molecule has an area of positive charge separated from an area of negative charge, it is called a *polar* molecule.

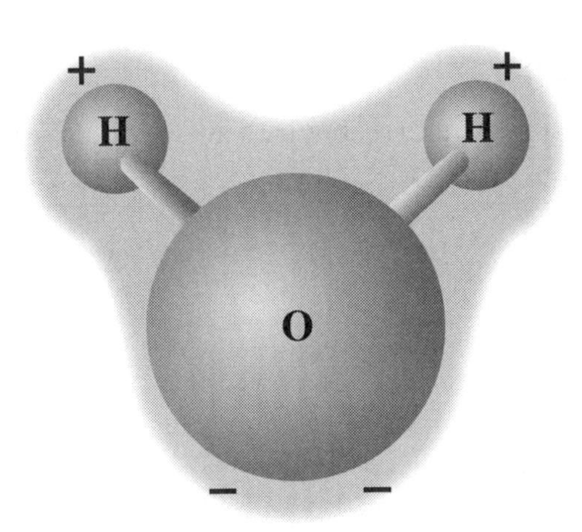

Polar water molecules interact with other polar or charged particles

Because of their positive and negative regions, water molecules are attracted to each other and to other particles that are charged or have polar areas. This behavior accounts for many of water's unique characteristics.

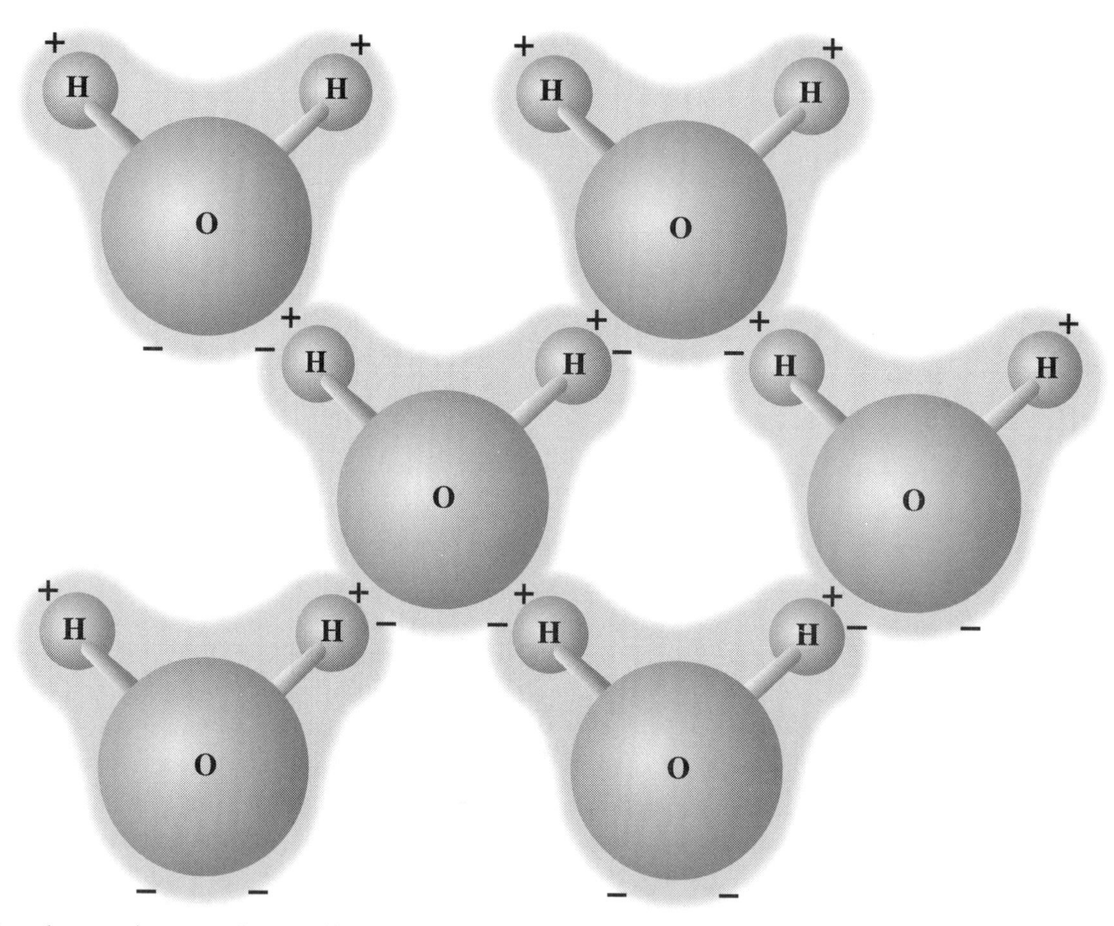

Many of the observations students will make in the activities in *Inquiry in Action* are a result of the characteristics of water. Water's ability to dissolve substances, its changes in state between gas, liquid, and solid, its role in making acids and bases, and its influence on chemical reactions are all related to the nature of the water molecule.

This first investigation can serve several purposes if it is done near the beginning of a physical science unit on matter. Students will see that the color on the outside of an M&M dissolves in water. As students observe that all M&Ms, regardless of their color, behave in the same way when placed in water, they will have discovered a *property* of M&Ms. This operational definition of "property," in the context of the M&M activity, can lay the foundation for generalizing the meaning of property to other objects, materials, and substances. After students observe one M&M in room temperature water, they may wonder if the color moves differently when two or more M&Ms are used. Students may not recognize it as such, but asking a question about the *number* of M&Ms means that they have identified a *variable* in the activity. Again, this operational definition of "variable" as something that might affect the way the M&M colors move can eventually be generalized to a broader understanding of variables as applied to any science experiment. Also, the formulation of a question such as "What would happen to the colors if we used two or more M&Ms?" is a good starting point for developing the ability to ask a question that can be answered in a scientific way. A well-formulated question can be very helpful in identifying variables and eventually designing a valid investigation.

Activity 1.1—One M&M in water

When an M&M is placed in water, the colored coating on the M&M dissolves into the water in a relatively circular pattern around the M&M. The color comes off the M&M because the coloring used is *soluble* in water. This means that the water molecules and the molecules that make up the coloring have an attraction for each other and mix together. The color dissolves in a uniform way around the entire M&M because the M&M is exposed to about the same amount of water all around its surface.

Activity 1.2—Racing M&M colors

It is possible that some colors of M&Ms are more soluble in water than others. The more soluble colors would dissolve into the water more quickly and make a larger circle of color faster. There may be a problem in judging which color moves faster since it may be more difficult to see certain colors than others. So even if yellow, for instance, makes a larger circle than brown in a given amount of time, it may not be as easy to see the yellow so it may appear that brown has gone further.

Activity 1.3a—Two M&Ms, and Activity 1.3b—M&Ms in different arrangements

When the colors from two or more M&Ms come together, the colors seem to form a kind of line or barrier and the colors do not readily mix. One reasonable explanation for this is that the color and the sugar from each M&M dissolve pretty well in water but don't dissolve as well in each other. You can show this by using food coloring and corn syrup to make a sample of blue and a sample of red corn syrup. Add water to a plate and place a few drops of red corn syrup together near a few drops of blue corn syrup. Watch what happens when the red and blue corn syrup meet.

Activity 1.4—M&M colors in different temperatures

The M&M color dissolves in hot water faster than it does in room-temperature or cold water. The water molecules in hot water move faster and make more contacts with the color on the M&Ms, resulting in more dissolving. This is true for many substances but not all. Common table salt does not dissolve much faster in hot tap water than in room temperature water. You could have your students design an experiment to see if this is true.

Activity 1.5—M&M colors in different liquids

The reason the M&M color comes off the M&M differently in different liquids is probably a result of several factors. The solubility of a substance depends on the *solvent* being used. The solvent is the part that is in greater amount and doing the dissolving. For example, substances that are very soluble in water may not be so soluble in alcohol. The presence of other solutes can also influence the solubility of substances.

Activity 1.1
One M&M in water

Question to investigate

What happens when one M&M is placed in water?

In this activity, students will see that when an M&M is placed in water, the colored coating dissolves and spreads in a circular pattern around the M&M. The way the color moves in water is a *property* of M&Ms. When students compare their results, they will discover that every color of M&M moves in a similar way. This property of the colored coatings provides a basis for students to ask questions and investigate the property further.

1. Have students describe some of the properties of M&Ms.

Pass out an M&M to each student or group of students. Then, ask the students questions like the following:

- How would you describe the size, shape, color, texture, etc., of the outside of an M&M?
- Have students break open an M&M so that they can observe the inside.
- How would you describe the properties of the inside of an M&M?
- Has anyone ever noticed what happens when the colored coating of an M&M gets wet?

Explain to students that their descriptions of M&Ms are all *properties* of M&Ms.

The way the color comes off an M&M when it gets wet, or how it changes when something else is done to it, are also properties of M&Ms.

- What do you think might happen to the color of an M&M if you accidentally dropped it in a puddle of water?

Tell students that in this activity they will see what happens to the color of an M&M when it is placed in water.

2. Have students place an M&M in a dish of water.

Giving students an opportunity to observe an M&M in water will give them the context and motivation to want to find out more about how M&M colors move in water. From this experience, you can get them to ask questions that they can investigate.

Procedure

1. Fill a 3½-ounce cup with room-temperature water. Pour the water into a white plastic or foam plate so that it just covers the bottom. Add more water if necessary.

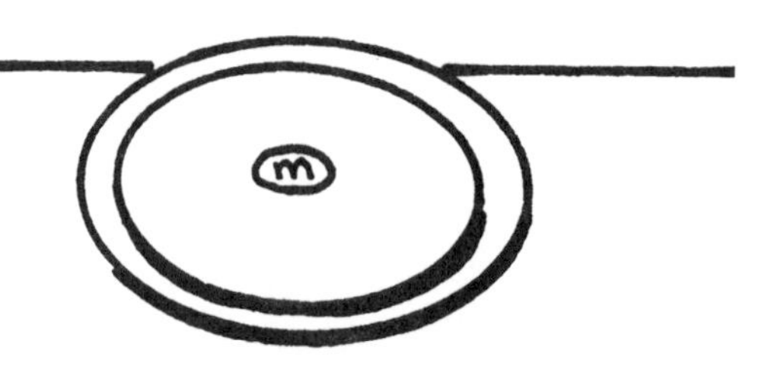

2. Place one M&M in the center of the plate. Be careful to keep the water and M&M as still as possible. Observe for about 1 minute.

3. Have students compare their results.

Ask students what they notice about the movement of the color from their M&M.

Expected results: **Each colored coating of M&M will dissolve in a circular pattern around the M&M. If anyone notices differences such as "the color moved over to one side more than the other," check to see that the plate is level.** Point out to students that since the colored coating on M&Ms moves in a similar pattern each time they are placed in water, this movement is a *property* of the M&M coating.

Empty the plate of water and candy into a bucket, bowl, or sink. Dry the plate with a paper towel.

4. Ask students what else they could investigate about the way M&M colors move in water.

Remind students that they have tested *1 M&M* of a certain *color* in a plate of *water* that is at *room-temperature*. The number of M&Ms, the color, the type of liquid they are placed in, and the temperature of the liquid are things that can be changed to do new experiments. As a whole class, have students brainstorm a list of questions they would like to investigate with new experiments. Pass out *Activity sheet 1.1—One M&M in water* on page 32 to students so that they can select a question and plan how they might investigate it.

Sample procedures for the following questions are included in this investigation. Feel free to have students ask questions about and investigate other questions that interest them. Students will record questions like the following, as well as the results of their experiments on *Activity sheet 1.2, 1.3, 1.4, 1.5—Investigating questions about M&Ms in water* on page 33.

- Do some M&M colors move faster in water than others? (Activity 1.2, page 24)
- What would happen to the movement of the color of one M&M if we added another M&M to the plate? (Activity 1.3a, page 26)
- What would happen if we put 2 or more M&Ms in different positions in the water? (Activity 1.3b, page 27)
- Will hot or cold water make the M&M color move differently than it does in room-temperature water? (Activity 1.4, page 28)
- Does putting the M&Ms in a liquid other than water affect the movement of the color? (Activity 1.5, page 30)

Activity 1.2
Racing M&M colors

Question to investigate

Do some M&M colors move faster in water than others?

Students will identify and control variables as they help design an investigation to find out if all colors of M&Ms move at the same speed. Students may discover that they all move at about the same speed or that some appear to move a little slower or faster than others. Regardless of their results, it is important that students begin to understand the importance of conducting a fair test.

1. Ask students how they might design an experiment to answer the question.

In their groups, have students quickly brainstorm ways they could investigate the question. As you go around to groups and listen to their discussions, ask them what they are doing to keep the test as fair as possible. Students should have a plan to control variables such as type of plate, amount of water, temperature of water, how and where the M&Ms are placed on the plate, and the amount of time the M&Ms are in the water.

As a whole class, share group plans, drawing attention to methods of controlling variables. Tell students that for a scientific investigation to be valid or fair, all variables need to be kept the same except for the one being tested: In this case, it is the color of the M&M.

If the groups had good experimental designs, you may choose to have them follow their own procedures. If you feel that groups need more guidance, you could have a whole class discussion to design an experiment that all groups will do. Either way, pass out *Activity sheet 1.2, 1.3, 1.4, 1.5—Investigating questions about M&Ms in water* on page 33, so that students can record the question they are investigating and their observations. They should keep this activity sheet for the remaining activities in this investigation.

2. Have students compare the speed of M&M colors.

The following procedure is an example of one way students might investigate the speed of M&M colors in water. It is best to have students participate as much as possible in designing the investigation. The way the students in your class design and conduct the investigation may differ from the following procedure. This procedure is provided so that you have an example of one experimental design that addresses the question and is a fair test.

Procedure

1. Draw a target on each plate so that you can better compare the speed that the colors move out from the M&Ms. Use a permanent marker to trace around the top of a 3½-ounce cup to draw a circle in the center of each of the 7 plates. Turn the cup over and use the bottom to draw a smaller circle inside the larger one. Make a dot in the center of the circles.

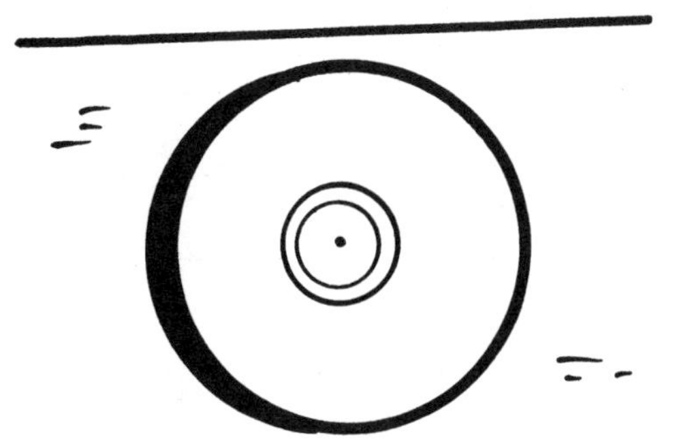

2. Fill a small bathroom-size cup (about 3½ ounces) with room-temperature water. Pour the water into each Styrofoam plate so that it just covers the bottom. Add more water if necessary.

3. With the help of your lab partners, place a different-colored M&M in the center of each plate at the same time. Observe for 1 minute.

3. Have students report their results.

Ask students questions like the following:

- Does one color seem to move faster than the others?
- Did other groups have similar results?
- Is there enough evidence from your experiments to conclude that a particular color of M&M moves faster in water than the others?

Expected results: **Groups may have different results even if they did their best to conduct fair tests.** There are several possible explanations:

- the experimental design is not sensitive enough to discern minor speed differences,
- there was human error in conducting the experiment, or
- the speed at which a particular color moves in water is not consistent.

Empty each plate of water and M&M into a bucket, bowl, or sink. Dry the plates with a paper towel.

Activity 1.3a
Two M&Ms

Question to investigate

What would happen to the movement of the color of one M&M if we added another M&M to the plate?

When students do this activity, they may be surprised to find that the addition of another M&M changes the way the color from the first M&M moves.

1. As a class, design and conduct an experiment to answer this question.

Ask students how they might do the experiment. When you decide on a class plan, have groups conduct the experiment. They should record their questions and observations on *Activity sheet 1.2, 1.3, 1.4, 1.5—Investigating questions about M&Ms in water* on page 33.

The procedure written below is an example of one way students can see if another M&M has any affect on the flow of the color from the first M&M. This procedure uses two identical plates of water. One has one M&M while the other has one M&M of the same color plus an additional M&M. The plate with only one M&M is the *control*. The purpose of the control is to help students compare the way color moves from one M&M to the way color moves when another M&M is added. Since students have already seen the movement of color from one M&M, they probably will not think of setting up one M&M in water again. Suggest that students include a control in the design of their experiments and explain why it is useful.

Procedure

1. Fill a small bathroom-size cup (about 3½ ounces) with room-temperature water. Pour this amount of water into each of two Styrofoam plates. Place the same color M&M in the center of each plate.

2. Once the color begins to move, place a different color M&M about 2–3 centimeters from the M&M in one of the plates. Wait about 1 minute.

2. Have students share their observations.

Expected results: **The colors will flow in a circular pattern around each M&M until the color approaches the color from the other M&M. The colors will not blend in the middle of the plate, but will form a distinct line where the colors meet. Some students may notice that certain colors slowly mix or flow over or under other colors. In these areas, when looking from the top, the colors appear blended.**

Empty the plate of water and M&M into a bucket, bowl, or sink. Dry the plates with a paper towel.

Activity 1.3b
M&Ms in different arrangements

Question to investigate

What would happen if we put 2 or more M&Ms in different positions in the water?

Each group may have several different arrangements of M&Ms that they would like to try.

1. Encourage students to experiment with placing M&Ms in different arrangements in water.

Ask students to draw or suggest arrangements of M&Ms in water that they would like to try. You could then ask students to predict what pattern they think the M&M colors will make in the water with each arrangement. Have small groups decide which arrangements they would like to try, test at least one of them, and then record their results. Students should record their question and observations on *Activity sheet 1.2, 1.3, 1.4, 1.5—Investigating questions about M&Ms in water* on page 33. If they would like to test more than one arrangement of M&Ms, they will likely need another activity sheet.

Students may try arrangements like the following:

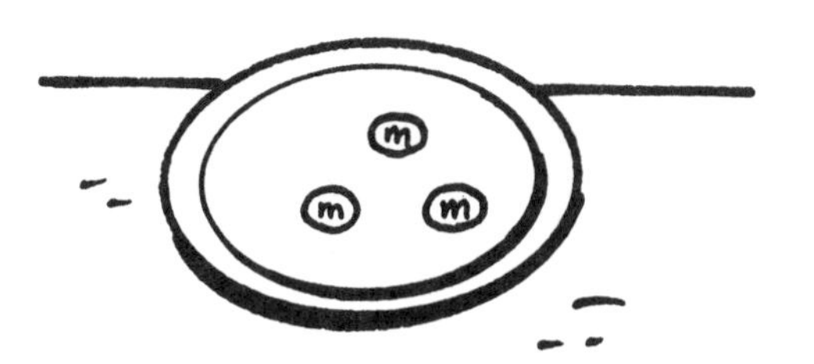

- What would happen if you spread 3 M&Ms in a triangle shape?
  ...4 in a square shape?
  ...4 in a square shape with 1 in the middle?

- What would happen if you placed 3 M&Ms in a row in a dish of water?
  4? 5?

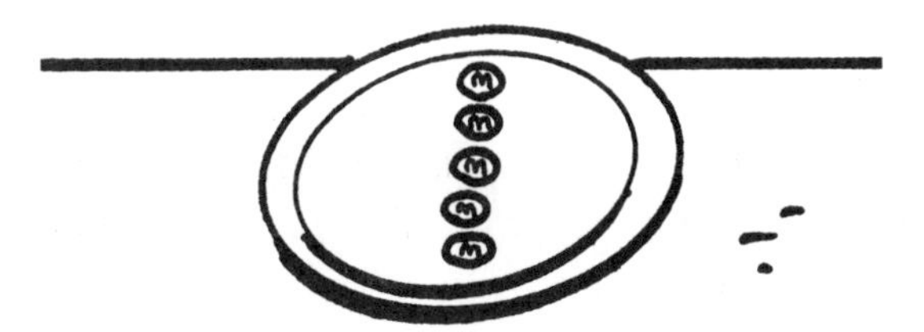

- What would happen if you placed 3 M&Ms in a row along the edge of a plate?
  4? 5?

2. Have students communicate their results.

Once students have completed their experiments, have a class discussion in which students share what they tried and their observations. Have groups who tried similar arrangements compare their observations.

Activity 1.4
M&M colors in different temperatures

Question to investigate

Will hot or cold water make the M&M color move differently than it does in room-temperature water?

In this activity, you will help students design an experiment to answer the question. Be sure to have students think about the variables they will need to control.

1. Have groups suggest experimental designs.

In small groups, have students quickly brainstorm ways they could investigate the question. As you go around to groups and listen to their discussions, be sure that students are thinking about variables. They should consider variables such as the kind of plate, the amount of water in each plate, the color of the M&Ms, and the placement of the M&Ms at the same time in each plate. All these variables should be kept the same. Students should realize that the only variable that should be changed is the temperature of the water.

2. As a whole class, finalize the experimental design.

Have students share their plans with the whole class. Some groups may have planned to test M&Ms in cold and room temperature water, or hot and room temperature water. Other groups may suggest placing an M&M in hot, cold, and room-temperature water. Encourage all groups to test an M&M in all three temperatures of water. The room-temperature water is used as a control to see the effect of the hot and cold water on the movement of the M&M color.

Students should record the question and their observations from this activity on *Activity sheet 1.2, 1.3, 1.4, 1.5—Investigating questions about M&Ms in water* on page 33.

3. Have groups conduct the experiment.

Your discussions with students have probably resulted in a procedure similar to the following. This procedure is provided so that you have an example of one experimental design that addresses the question and is a fair test.

Procedure

1. Use a permanent marker to trace around the top of a 3½-ounce cup to draw a circle in the center of each plate. Turn the cup over and use the bottom to draw a smaller circle inside the larger one. Make a dot in the center of the smaller circle.

2. Fill a small bathroom-size cup with room-temperature water and pour the water into a Styrofoam plate. Similarly, add hot water and cold water to each of 2 Styrofoam plates.

3. At the same time, place the *same*-colored M&M in the center of each plate.
   Wait for 1 minute and observe.

4. Have students share their observations.

Ask students if they noticed a difference in the movement of the color in the different temperatures of water.

Expected results: **The M&M color dissolved and spread out fastest in the hot water. The color moves similarly in cold and room-temperature water, but a little slower in the colder water.**

Activity 1.5
M&M colors in different liquids

Question to investigate

Does putting the M&Ms in a liquid other than water affect the movement of the color?

The design of this experiment is similar to Activity 1.4. In both activities it is important to have students identify and control variables and to recognize the usefulness of a control.

1. Have students help design an experiment to answer the question.

As a whole class, ask students for suggestions of ways they could find out if placing an M&M in different liquids, such as salt water or sugar water, might affect the movement of the M&M color. Be sure to have students think about the variables they will need to control. Students should consider variables such as the type of plate, the amount of liquid in each plate, the color of the M&Ms, and the placement of the M&Ms at the same time in each plate. All these variables should be kept the same. Students should realize that the only variable that should be changed is whether the liquid is salt water, sugar water, or fresh water. They also should realize that the plate of fresh water will serve as a control.

Students should record the question and their observations from this activity on *Activity sheet 1.2, 1.3, 1.4, 1.5—Investigating questions about M&Ms in water* on page 33.

2. Have small groups conduct the experiment.

Your discussions with students have probably resulted in a procedure similar to the following. This procedure is provided so that you have an example of one experimental design that addresses the question and is a fair test.

Procedure

1. Use a permanent marker to trace around the top of a 3½-ounce cup to draw a circle in the center of each plate. Turn the cup over and use the bottom to draw a smaller circle inside the larger one. Make a dot in the center of the circles.

2. Label 2 cups *salt* and *sugar*. Add 1 small bathroom-size cup of room-temperature water to each cup.

3. Add 4 teaspoons each of salt and sugar to their labeled cups. Stir until both are as dissolved as possible.

4. Pour the salt water into one plate, sugar water into another, and 1 small bathroom-size cup of room-temperature water into the third plate.

5. With the help of your lab partners, place the same-colored M&M in the center of each plate at the same time. Wait about 1 minute and observe.

3. Have students share their observations.

Have students describe the way the M&M color moved in the salt water, sugar water, and fresh water.

Expected results: **The M&M color spread out in the following order from most to least: water, sugar water, and salt water.**

Name:______________________________

Activity sheet 1.1

One M&M in water

After observing 1 M&M in a plate of water, think of some other experiments you could do with M&Ms in a plate of water. Brainstorm your ideas in the form of questions and write them in this box.

Choose one question that you think would be interesting to investigate. You should be able to investigate this question in your classroom. Write your question in this box.

Draw a picture of an M&M (or M&Ms) arranged in a plate that shows one way to investigate your question. Label your drawing if necessary.

Write what you would do to answer the question.

Name:______________________________

Activity sheet 1.2, 1.3, 1.4, 1.5

Investigating questions about M&Ms in water

Question to investigate "What would happen if…"	Draw what happened	Answer the question

Name:______________________________

Assessment rubric

Investigation 1—Scientific questions and their investigation

To earn a "B," a student must receive a "Very good" in each category.

	Very good	Satisfactory	Needs improvement
Activity sheet 1.1 **One M&M in water** Includes at least 5 questions Drawings reflect selected question Well-designed plan to answer the question	______	______	______
Activity sheet 1.2, 1.3, 1.4, 1.5 **Investigating questions about M&Ms in water** Records actual investigations Questions are well-written Uses clear drawings of experimental setup Answers questions based on observations	______	______	______
Investigative behaviors Contributes to brainstorming questions Conducts fair tests Participates in investigations Works cooperatively with group	______	______	______

To earn an "A," a student must also exhibit some of the following qualities throughout this investigation.

Develops reasonable well-written questions
Creates outstanding drawings and written explanations
Participates well in class discussions
Participates well in group work
Uses scientific thinking
Consistently exhibits exceptional thought and effort in tasks

Class discussion: Concept of variable and control

Discussion and home experiment

When students conducted the activities in Investigation 1, they began to think about the importance of *variables* in scientific investigations. For example, when investigating M&M colors in different temperatures, students were careful to control variables such as the amount of water, the color of the M&M, and when the M&M was placed in the center of each plate. The point was made that all of these factors that could affect the experiment are called *variables*. Since the concept of variable is an essential part of designing any scientific investigation, this class discussion and activity are intended to help students develop a better understanding of the concept of variable.

Almost any simple investigation can be used to explore the concept of variable. The following discussion and activity focus on a basic question that students often wonder about and that can be easily investigated.

Which freezes faster, fresh water or salt water?

In a class discussion, students will consider the variables involved in designing and conducting an experiment to answer this question. Students will then conduct their own experiment at home to answer the question. They will write about what they did, report their observations, and answer the question.

1. Ask students how they would design an experiment to answer the question: *Which freezes faster, fresh water or salt water?*

Discuss the following variables by asking questions such as those listed below. Your questioning strategy will lead this discussion. The suggested questions included in this discussion are meant to serve as a guide as you implement your own questioning strategy.

Type of cup

Should the type of cup used for the salt water and the fresh water be the same?

- Do you think that the type of cup used might have an effect on how fast the liquid in each cup freezes?
- If you put the fresh water in a thin metal cup and the salt water in a thick insulated foam cup, would that be a fair test? Why not?
- Since the type of cup might affect the experiment, do you think you should make sure the cups are the same?

Amount of fresh water and salt water

Should you put the same amount of salt water and fresh water in the cup?

- Do you think that the amount of salt water and fresh water used might have an effect on how fast each one freezes?
- If you used 1 drop of fresh water and 2 gallons of salt water, would that be a fair test? Why not?
- Since the amount of water might affect the experiment, do you think you should make sure the amounts are the same?

Temperature of fresh water and salt water

Should the temperature of the salt water and the fresh water be the same?

- Do you think that the temperature of the salt water and the fresh water used might have an effect on how fast each one freezes?
- If the fresh water was 100 °F when you put it in the freezer and the salt water was only 35 °F, would that be a fair test? Why not?
- Since the temperature of the water might affect the experiment, do you think you should make sure the temperatures are the same?

Temperature of freezer

Should you put the cups of salt water and fresh water in the same freezer?

- Do you think that the freezer used could have an effect on how fast each one freezes?
- If you put the fresh water in a freezer set at 32 °F and the salt water in a freezer set at 25 °F, would that be a fair test? Why not?
- Since the temperature of the freezer might affect the experiment, do you think you should make sure to place the cups in the same freezer?

2. Summarize the discussion and meaning of "variable."

Explain to students that all these factors—the type of cup, the amount of water, the temperature of the water, and the freezer used—that could affect how fast the fresh water and salt water freeze are called *variables*. The idea in a valid experiment, or a fair test, is to keep all variables the same for both cups except for the one variable you are experimenting with. In this experiment, the only variable that should be different between the 2 cups is the salt content—1 cup has fresh water and the other has salt water. Other than that, all variables should be the same for both cups.

3. Have students conduct this experiment at home.

Tell students about their home assignment.

- Design and conduct an experiment using materials that you have at home to answer the question: *Which freezes faster, fresh water or salt water?*
- Write what you did, your observations, and your answer to the question on *Activity sheet—Concept of variable and control* on page 39.

Teacher note: Have students use disposable plastic or paper cups, not glass. In order to have more consistent results, have students add 1 tablespoon of salt to ¼ cup of water to make the salt water.

4. When students have completed their experiments, compare experimental designs and results.

Discuss the common features of everyone's experiments and strategies for controlling the variables. Discuss the methods students used to determine when either sample was frozen.

A case study: Louis Pasteur and the theory of spontaneous generation

A historical account of identifying and controlling variables

Some people used to think that flies, worms, bacteria, and other unwanted organisms actually came from rotten food, liquid, or other substances. They thought that somehow the food actually *turned into* these organisms. This idea—that nonliving substances could turn into living organisms—is called *spontaneous generation*. *Spontaneous* means to happen suddenly without anyone or anything trying to make it happen. *Generation* means to come into being or to be born.

In the 1860s, the French scientist Louis Pasteur designed and conducted a scientific investigation to test whether the idea of spontaneous generation was true. Pasteur did not believe that food or drink could somehow turn into living organisms. Instead, he believed that the organisms came from somewhere else and got into the food from the air or in some other way. He felt it was important to design a fair experiment to test whether spontaneous generation was true or false.

Pasteur decided to concentrate on the problem of bacteria causing certain liquids to spoil. His question to investigate was "Do bacteria from the air cause food to spoil?"

In designing his experiment, Pasteur decided to use two containers of broth (like a clear soup). He knew that he needed to keep everything about these two containers exactly the same except for the one thing he was trying to test. Pasteur had to set up the experiment so that the only difference between the two containers was that bacteria could get into one container but not get into the other. Everything else about the containers needed to be the same: Both needed the same type of broth, both had to be open to the same air, and both needed to be exposed to the same light and temperature.

Pasteur predicted that the broth in which bacteria could enter from the outside would soon become filled with bacteria and would spoil. He also predicted that the broth that bacteria could *not* enter would not spontaneously produce bacteria but would remain clear and unspoiled. Pasteur believed that this experiment could show that the idea of spontaneous generation was not true.

Here's what he needed to do

Pasteur needed to figure out a way to let bacteria get into one container but not into the other. He could not simply leave one open and the other closed because then one container would be getting air and the other would not. This would be a difference between the two containers that Pasteur knew would make the experiment unfair. The only difference between them could be that bacteria could get into one but not the other. Pasteur needed to figure out a way to do this while leaving both containers open to the air.

Here's how he did it

Pasteur got two glass containers for holding the broth. One of the containers had a neck that went straight up and was open at the end. When air passed over the opening, bacteria in the air could fall down into the broth.

The other container had a *curved* neck that was open at the end. When air passed over this opening, bacteria would fall into the curve in the neck and become trapped, never able to reach the broth. Using this method, Pasteur found a way to expose both samples of broth to the air but allowed bacteria to get into the broth in only one.

Pasteur then put the same kind and same amount of broth into both containers. He heated each container, in the same way, at the same temperature, for the same length of time, to kill any bacteria that may have been in the broth already.

After only a few days, the broth in the straight-necked container, which bacteria could enter from the outside, was cloudy and spoiled. The broth in the curved-necked container that bacteria could not reach stayed clear.

The experiment proved that broth doesn't somehow spontaneously turn into bacteria on its own. Rather, for broth to spoil, bacteria need to get into it from the outside.

See the *Activity sheet—Louis Pasteur's famous experiment* on page 40.

Name:_______________________________

Activity sheet

Concept of variable and control

Which freezes faster, fresh water or salt water?

How did you set up your experiment?

____

____

____

____

____

____

____

____

____

Which variables did you keep the same so that your experiment was as fair as possible?

____

____

____

____

____

Which freezes faster, fresh water or salt water?

____

Name:______________________________

Activity sheet

Louis Pasteur's famous experiment

1. In Louis Pasteur's experiment to disprove the theory of spontaneous generation, what were some of the variables that he needed to keep the same so that his experiment was fair?

____

____

____

____

____

2. In Pasteur's experiment, what was the one variable that was different between the two containers?

____

____

3. If Pasteur wanted bacteria to get into one container of broth but not the other, why didn't he just leave one open and put a lid on the other?

____

____

____

____

4. In your own words, explain how the careful setup of Pasteur's experiment added evidence to the case that the theory of spontaneous generation might not be true.

____

____

____

____

Investigation 2.
Physical properties and physical change in solids

Summary

In this investigation, students will compare the properties of three different household crystals to the properties of an unknown crystal. This unknown crystal is chemically the same as one of the known crystals, but does not appear the same. Students will conduct tests for appearance, hardness, solubility, and recrystallization to help them identify the unknown crystal. The activities will emphasize identifying and controlling variables to design fair tests, making observations, and analyzing results.

Objective

Through testing different household crystals, students will develop an understanding of the meaning of *characteristic properties of substances*. Students will identify an unknown crystal by comparing its characteristic physical properties and the way it undergoes physical change with those of three known crystals. Students will identify possible variables and suggest ways to control them as they help design valid scientific investigations.

Assessment

The assessment rubric *Physical properties and physical change in solids* on page 64 is included so that you can assess and document student progress throughout the investigation. The abilities and understandings demonstrated by students and recorded on the rubric include the following: Students will design and conduct a series of tests exploring some physical properties of crystals. They will control variables to make each test as fair as possible. They will also understand that in order to measure equal amounts of crystals, the substances must be weighed instead of measured by volume. After completing the tests that students help design, they should be able to correctly identify the unknown crystal. Investigative behaviors observed as students plan and conduct their investigations, communicate their observations, and work with their groups are also recorded on the rubric.

Relevant *National Science Education Standards*

Physical Science

K–4

Properties of Objects and Materials

Objects have many observable properties, including size, weight, shape, and color.

5–8

Properties and Changes of Properties in Matter

Substances have characteristic properties, such as density, boiling point, and solubility.

Science as Inquiry

K–4

Abilities Necessary to do Scientific Inquiry

Ask a question about objects.
Plan and conduct a simple investigation.
Employ simple equipment and tools to gather and extend the senses.
Use data to construct a reasonable explanation.
Communicate investigations and explanations.

Understandings about Scientific Inquiry

Scientific investigations involve asking and answering a question.
Types of investigations include describing objects…and doing a fair test.
Good explanations are based on evidence from investigations.

5–8

Abilities Necessary to do Scientific Inquiry

Identify questions that can be answered through scientific investigations.
Design and conduct a scientific investigation.
Use appropriate tools and techniques to gather, analyze, and interpret data.
Develop descriptions, explanations, predictions, and models using evidence.
Think critically and logically to make the relationships between evidence and explanations.
Communicate scientific procedures and explanations.

Understandings about Scientific Inquiry

Different kinds of questions suggest different kinds of scientific investigations.
Scientific explanations emphasize evidence and have logically consistent arguments.
Scientific investigations sometimes result in new ideas and phenomena for study or generate new procedures for an investigation. These can lead to new investigations.

How this investigation relates to the *Standards*

This activity is designed to achieve two related goals of the *Standards*. One is for students to discover that substances have characteristic properties and undergo characteristic physical change. The other goal is for students to begin to develop the skills of designing and conducting fair tests to discover these properties. Since these physical properties are characteristic to each substance, they can be used to help identify an unknown substance. Students will discover that in order to compare the properties of different substances, each substance needs to be tested in the same way. These ideas concerning identifying and controlling variables are central to the design of any scientific investigation.

Materials chart

2.1 Appearance test
2.2 Crushing test
2.3 Solubility test
2.4 Recrystallization test

Each group will need

	Activities 2.1	2.2	2. 3	2.4
salt in cup	•	•	•	•
Epsom salt in cup	•	•	•	•
MSG (Accent®) in cup	•	•	•	•
kosher salt in cup (unknown)	•	•	•	•
black construction paper, ½ piece	•	•		*
magnifying glass	•			
masking tape	•	•	•	•
pen/pencil	•	•	•	•
ruler			•	
clear plastic cups			4	4
small cups (3½-ounce)			5	5
plastic teaspoon	•	•	•	•
hot water in source cup			•	•
paper clips			•	
cotton swabs				*

Notes about the materials

Standard metal paper clips weigh about 0.4–0.5 grams each. Depending on your paperclips, students should use 10 to measure about 4 or 5 grams of each crystal.

Activity 2.4 uses the same solutions from activity 2.3.

* The black construction paper and cotton swabs are optional.

Teacher preparation

Teacher will need for the demonstrations

2a Identifying the variables in a solubility test
2b Measuring equal amounts of crystals for the solubility test

	Demo 2a
overhead projector	•
transparency	•
transparency marker	•
clear plastic cups	2
bucket or sink	•
salt	•
sugar	•
room-temp water	•
teaspoon measure	•
scale to weigh 5 grams	•

	Demo 2b
primary balance scale	•
clear plastic cups	2
zip-closing plastic bag	•
ball-shaped cereal	•

Activity sheets

Copy the following activity sheets for this investigation and distribute them as specified in the activities.

Activity sheet 2.1, 2.2, 2.3, and 2.4 — page 61
Using physical change to identify an unknown solid

Copy one per student.

This activity sheet will be used for each of the activities in this investigation. As students conduct each test, they will circle all possible identities for the unknown and briefly explain why each crystal may or may not be the unknown.

Demonstration sheet 2b — page 62
Measuring equal amounts of crystals for the solubility test

Copy one per student.

The purpose of this activity sheet is to make sure that students understand the importance of using equal amounts of crystals for a test like the solubility test. Students should begin to realize as they watch the demonstration that weight is the best way to determine equal amounts of the crystals.

Activity sheet 2.3 — page 62
Solubility test

Copy one per student.

Students will record the results of their solubility test on this activity sheet. They will also apply their learning from the demonstration as they explain why it is best to weigh the crystals to get equal amounts as they did for this test.

Science background information

The physical properties of a substance are those characteristics of the substance that can be described without changing the substance's identity. For solids, some physical properties are shape, color, size, and texture. Some other physical properties of solids that are not as readily observable are density and hardness.

A concept related to physical properties is physical change. Physical change describes a type of change in a substance that does not change the fundamental identity of the substance. One physical change for a solid is melting to change from a solid to a liquid. Another is breaking apart and dissolving in a liquid to become part of a solution. In both cases, the solid changes its form or size but does not change its identity. The physical properties of a solid and the way it undergoes physical change are characteristic properties of that solid and can be used to distinguish it from other solids.

Activity 2.1—Appearance test

There are two main reasons why the various types of crystals look different from one another. One reason is the different atoms and molecules the crystals are made of and the way these atoms and molecules are bonded together. This atomic and molecular structure of the crystals affects their overall shape, color, texture, and other features. The other factor that can affect the appearance of the crystals is the way they are processed and packaged for sale. The table salt (sodium chloride) and the kosher salt (sodium chloride) are chemically the same but look different because of the way they are processed. Both salts are produced by pumping water into rock salt deposits and then collecting the salty water and evaporating it. To make kosher salt, the salty water is continuously raked during the evaporation process, resulting in a less uniform and flakier salt. Also, table salt has magnesium carbonate added to it as an anti-caking agent as well as iodine to help prevent thyroid gland problems, whereas kosher salt does not contain these additives.

Activity 2.2—Crushing test

The hardness of the crystals is mostly dependent on their atomic and molecular structure. But, as mentioned above, the processing of the crystals may also have an impact on their properties, including their hardness. This hardness test is not the classic hardness test that geologists use to help them identify minerals. It is a much more subjective test that has several variables that are difficult to control, which makes it a fairly unreliable test. It is included as an activity mainly because students may suggest crushing the crystals as a way to help identify them. Also, a discussion that identifies variables and considers why they are hard to control can help students better appreciate the issues involved in designing a valid experiment. The main variables that are difficult to control here are the crystal size and the force applied to the crystals. After doing the test and discussing the problems with certain variables, students will probably not be able to conclusively identify the unknown or even eliminate any crystals based on this test.

Demonstration 2b—Measuring equal amounts of crystals for the solubility test

To compare different types of crystals based on a dissolving test, it is important to start with the same amount of each type of crystal. This is not as easy as simply measuring out a level teaspoon of each. One problem is that the size and shape of the crystals will affect how much room they take up in the spoon, and therefore how much crystal will make a level teaspoon. If one type is very small and well-packed in the spoon, and the other is larger and packed more loosely, it is likely that a level teaspoon of each could contain a very different amount of crystal.

But even if you had two types of crystals with the exact same size and shape, they might have very different densities. Density is the amount of substance packed into a given space. Even if two types of crystals were the same size and shape, but one was more dense than the other, the one that was more dense would actually have more substance in each of its crystals. It wouldn't be fair to test a level spoonful of each because you would be testing more of the higher-density substance than the lower-density one.

The only fair way to measure equal amounts of crystal is to weigh them. If the two samples weigh the same then you know you are dissolving the same amount of each.

Activity 2.3—Solubility test

The reason why the different crystals dissolve differently has to do with the way water molecules interact with the particles of the crystals they are in contact with. The amount of attraction that the water molecules have on the particles of the substance compared to the attraction these particles of the substance have for each other determines if and how quickly the water can dissolve the substance. If students use the same amount of each crystal, the same amount and temperature water, and swirl in the same way, for the same length of time, this should be a pretty fair test of the differences in solubility between the crystals. But there is still one variable that has not been controlled, and that is the size of the crystals. The more surface area that is in contact with the water, the faster the crystal should dissolve. So for a given amount of crystal, the smaller it is crushed up to begin with, exposing a lot of surface area, the faster it will dissolve. Technically, to help reduce the effect of this variable, all crystal samples would have to be crushed to the same size. We do not recommend this step because of logistical difficulties in crushing the crystals to the same size.

Activity 2.4—Recrystallization test

The recrystallization test is like a reverse dissolving test. Instead of adding water, which breaks the crystals apart, water evaporates, allowing the crystals to reform. As the water evaporates, the concentration of the crystal molecules increases. This increase in concentration allows more molecules to associate with one another and reform bonds to crystallize again.

How scientists identify unknowns

In each of the investigations, *Physical properties and physical change in solids*, *Physical properties and physical change in liquids*, and *Chemical change*, students will develop tests that they will first apply to a set of known substances and then to an unknown to attempt to identify it. This is very similar to the process that scientists actually use to identify unknown materials and substances.

In geology, for example, tests have been developed and used on known rock and mineral specimens that show certain characteristic properties of the specimen being tested. These tests were worked out by trying many different types of tests over a period of time until certain ones were found to give useful and reliable information on a consistent basis. Some tests include trying to scratch the specimen with materials of different hardness to see how hard the rock or mineral is. Another involves scraping the specimen on a white surface to see what color streak it makes. Another test entails dripping one or more chemicals on the specimen to check for a certain reaction. Geologists have recorded the results from these tests on many known rocks and minerals. When geologists find a new rock or mineral, they do these same tests to see if the new sample has similar properties to the known specimens. Using this method, geologists can begin to identify the new rock or mineral.

Another example is in biology. Biologists also have a set of tests that they have developed over the years on known bacteria. These tests show certain characteristic properties of the different bacteria. One of these tests is an appearance test under a microscope to look at the size and shape of the bacteria and whether the bacteria cells are individual, clustered, or in a chain. Biologists have also discovered that different bacteria look different when certain stains are used to color them. Microbiologists who study bacteria have developed many staining tests to help identify bacteria. They have recorded the results of these tests on many different known bacteria. When unfamiliar bacteria are found, these same tests are applied and compared to the results from the known bacteria. This process helps the scientist begin to identify the unknown bacteria.

An obvious example, but a bit unsavory, is the identification of illegal drugs or their components. Again, scientists have developed a series of tests on known samples of illegal drugs and their components. The results of these tests have been recorded and can be used by scientists who need to try to tell whether an unknown substance is illegal. When a substance is thought to be illegal, the series of tests is applied and the results are compared to the results from the known samples. This process can help tell scientists whether the unknown is an illegal substance.

Activity 2.1
Appearance test

Question to investigate

Can you identify an unknown crystal by comparing its appearance to other known crystals?

In this activity, students will look at three known crystals. They will then be given an unknown crystal that they will try to identify from appearance alone. Students will discover that this test does not give enough information to identify the unknown.

1. Have students carefully look at three known crystals.

Have students look at small samples of Epsom salt, table salt, and MSG on a piece of black construction paper. They can look at the crystals and touch them but they should not taste the crystals. Students should describe physical properties such as size, shape, color, texture, whether shiny or dull, transparent or opaque. After allowing students to examine the crystals closely, ask them about any similarities and differences they notice among the crystals. Then ask them if they think they could identify a new sample of one of these crystals just by looking at it.

2. Have students examine an "unknown crystal" and try to identify it.

Give students a sample of the unknown, which is coarse kosher salt. It is chemically the same as table salt, but looks different. Don't reveal the identity of the unknown until students have successfully identified it after a series of tests. Since it looks different from any of the three samples, students should be unable to identify it from an appearance test alone. Give students *Activity sheet 2.1, 2.2, 2.3, 2.4—Using physical change to identify an unknown solid* on page 61, so that they can explain the possible identity of the unknown based on their initial observations of each crystal. Students should keep this activity sheet to record the possible identity of the unknown based on each test.

Expected results: **The unknown will not look much like any of the other crystals.**

3. Ask students for suggestions of a test they could do to help identify this "unknown crystal."

Brainstorm a list as a whole class. Students may suggest weighing the crystals, seeing if they dissolve in water, crushing them, or some other ideas.

Activity 2.2
Crushing test

Question to investigate

Can you identify the unknown crystal by crushing crystals and comparing them?

If students suggest crushing the crystals to compare them, help students design a procedure they can try. While a crushing test will probably not give conclusive results, it is a good opportunity to encourage students to design investigations based on their own questions. With this activity, you and your students can also discuss variables and why they are sometimes difficult to control.

1. Have students help design a fair test.

If you and the students decide on a crushing test, lead a class discussion where students suggest ways to compare the "crushability" of the crystals. Important considerations to try to elicit from students involve controlling variables such as using the same object to crush each pile of crystals, and trying to use the same amount of force for the same length of time. Discuss the importance of keeping variables the same in a scientific investigation so that the test is fair. Students should also consider the importance of labeling each sample and keeping the samples separated on the black paper. As a class, decide on the materials and procedure the groups will follow. You and the students can, of course, decide to use a can, rolling pin, or any other safe object and safe method to apply a consistent amount of force to the crystals in the same way. The procedure below is just one possible experimental design. Be sure students wear safety goggles when crushing the crystals.

2. Have students conduct the experiment.

Procedure

1. Use masking tape and a pen to make a small label for each of the 4 crystals. Place the labels on 4 different areas on the black paper.

2. Spread a little of each of the 4 crystals in its labeled area on the black paper.
3. Use your thumb in the bowl of a plastic spoon to press down on each pile of crystals, as shown. Rock the bowl of the spoon back and forth to help crush the crystals.
4. Listen closely to the sound the crystals make as they break. Be aware of any difference between the way the crystals feel when they break.

3. Have students share and interpret their results.

Have students record their ideas about the identity of the unknown on *Activity sheet 2.1, 2.2, 2.3, 2.4—Using physical change to identify an unknown solid* on page 61.

Ask the following questions:

- Were there any crystals that were so different from the unknown that they could be eliminated?
- Were there any crystals that were similar enough to the unknown that it might be the unknown?

Expected results: **Although students probably cannot identify the unknown based on their crushing test, they may get a little information about what the unknown could be. Using the procedure described above, MSG makes well-defined cracks. Epsom salt, table salt, and the unknown all make cracking sounds that are less sharp and distinct than the MSG.**

Ask students if comparing the sound or feel of crushing crystals is the best way to identify the unknown. Students could conclude that the crushing test is not the best test because the crystals sound or feel too similar, and there were variables like the size of the crystals and the amount of force used that were difficult to control.

Demonstration 2a
Identifying the variables in a solubility test

Question to investigate

Do some of the crystals dissolve more or less than others?

Solubility, or whether and how well a substance dissolves, is a physical property of a substance. As a demonstration, you will place two types of crystals in water and compare the extent to which they dissolve. Since more of one crystal dissolves than the other, students will see that a solubility test might help tell the difference between types of crystals and help identify the unknown crystal. Your demonstration should start students thinking about the variables that they will need to control when they do their own solubility test on all of the crystals.

1. Have students identify variables in a solubility test.

Students may have thought of dissolving the crystals to help identify the unknown. If not, suggest that dissolving crystals may reveal more about the identity of the unknown than the appearance test and crushing test did. Tell students that salt and sugar are both crystals that dissolve in water.

Ask students the following types of questions to help them identify the variables in this demonstration:

- Should the cups be the same? Would it be fair if I used a foam cup for one and a glass cup for the other?
- In order to have a fair comparison, should we use the same amount of water in both cups when we try to dissolve the crystals?
- What about the temperature of the water? Why is that important?
- Should we use the *same* amount of each crystal? Why?

2. As a demonstration, dissolve salt and sugar in water.

Teacher preparation: Before this lesson, measure 5 grams each of salt and sugar. If you do not have a scale, make one according to the directions in Activity 2.3. Ten paper clips weigh between 4 and 5 grams. Place a transparency on the overhead. On the transparency, label one area salt and another sugar.

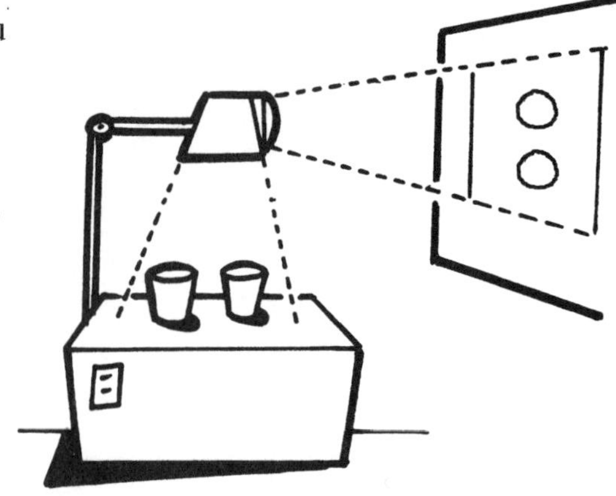

Follow the procedure below while pointing out to students that you are using the same type of cup, the same amount of water at the same temperature, and the same amount of crystal, swirled in the same way for the same length of time.

Procedure

1. Using identical clear plastic cups, place 1 teaspoon of hot tap water into each cup and place them on the overhead.

2. Place equal preweighed samples (about 5 grams) of salt and sugar into the cups of water *at the same time*. Swirl each cup at the same time and in the same way for about 20 seconds. Ask students if one substance seems to dissolve more than the other. Swirl again for 20 seconds and observe and then for 20 more seconds and have students make their final observations.

Expected results: **Much more of the sugar will dissolve than the salt.** Since more sugar dissolves in water than salt, a dissolving test is one way to tell salt and sugar apart.

3. Discuss the results of the demonstration and introduce the idea of "equal amounts."

Ask students if they think that they could use a dissolving test with their crystals to identify the unknown. Tell students that they will have a chance to dissolve their crystals in water the way that you did. Have them tell you what they should do to make the test as fair as possible. When students bring up using equal amounts of crystals, ask them how they could measure equal amounts. Students may suggest using a volume measure like a teaspoon or weighing the crystals. Tell students that they will explore which is the best method to measure "equal amounts" of the crystals.

Demonstration 2b
Measuring equal amounts of crystals for the solubility test

Question to investigate

What is the best way to measure "equal amounts" of crystals?

If you ask students how they could measure the same amount of each crystal, they may suggest using a volume measure like a teaspoon or weighing the crystals. In this demonstration, students will discover that weighing the crystals is the best way to measure "equal amounts." In this demonstration you will show students that weight rather than volume is the best measure of equal amounts.

Teacher preparation: Before this demonstration, place the exact same number of cereal balls into two identical clear plastic cups. Each cup should look like it is filled to the same height (about ¼ full).

1. Crush cereal to show students that the same amount of cereal may look different.

Procedure

1. Hold up two clear plastic cups, each containing the same number of pieces of cereal. Tell students that you have placed exactly the same number of cereal balls into each cup. Ask students if the amount of cereal in each cup looks about the same. Students should agree that the amount of cereal in each cup appears to be the same.

2. Ask students how the height of the cereal in the cup will change if you smash the cereal in one of the cups. Students will probably suggest that the smashed cereal will not take up as much room in the cup. Pour the cereal from one of the cups into a zip-closing plastic bag. Seal the bag and smash the contents thoroughly with the back of a spoon until the cereal looks like a powder.

3. Pour the pulverized cereal back into the cup. Hold both cups up and ask students if there is the same amount of cereal in each cup. Point out that although the pulverized cereal takes up less space, it is still the same amount of cereal as before since no cereal was added or removed.

2. Prove that the amount of cereal in each cup is the same.

Now that the two samples look different, ask students how they could show that the amount of cereal in each of the two cups is actually the same. Students might suggest weighing the cups. Place both cups on a balance to show that they weigh the same and to confirm that the cups contain the same amount of cereal.

Expected results: **The cups should balance on the scale.**

3. Conclude that, in order to measure "equal amounts," it is better to weigh substances than to measure by using volume.

Tell students that in the solubility test that they will do, they will need to measure equal amounts of four different crystals. Ask students what is the best way to measure "equal amounts." Students should realize that weighing is a better method to use than measuring by volume. In order to make sure that all students understand the important concept demonstrated, have them answer the questions on *Demonstration sheet 2b—Measuring equal amounts of crystals for the solubility test* on page 62.

Activity 2.3
Solubility test

Question to investigate

Can you identify the unknown crystal by the amount that dissolves in water?

In this activity, students will use solubility to distinguish between different crystals to help identify the unknown. As students design this experiment, they will refer back to the demonstration and discussion about controlling variables for a solubility test. The results of this test will most likely allow students to eliminate at least one of the crystals as the possible unknown.

1. Have students weigh equal amounts of the crystals.

The amount of crystal and water used in this solubility test is specific and should be used because it gives clear results. The way you choose to weigh equal amounts of each crystal may vary. You may use any scale that students are familiar with that can weigh about 4-5 grams.

Procedure

1. Use your masking tape and pen to label 4 small cups *salt*, *Epsom salt*, *MSG*, and *unknown*. Label 4 clear plastic cups in the same way. You should have 2 labeled cups for each type of crystal.

2. Tape the pencil down as shown. Roll 2 small pieces of tape so that the sticky side is out. Stick each piece of tape to the opposite end of the ruler. Place the small empty *salt* cup on 1 piece of tape so that the edge of the cup bottom is right at the end of the ruler. Place a small unlabeled cup on the other piece of tape in the same way.

3. Lay the ruler on the pencil so that it is as balanced as possible. Make a mark on the ruler at the point where it is balanced on the pencil. This is your *balance point*.

4. Carefully place 10 paper clips in the unlabeled cup. Slowly add salt to the *salt* cup until the cup with the paper clips just barely lifts from the table. Remove the *salt* cup from the ruler and set it aside.

5. Weigh the other 3 crystals in the same way so that you have equal amounts of all 4 crystals in their small labeled cups.

2. Discuss the variables that need to be controlled in the solubility test.

Ask students how they might mix the crystals into water to compare how they dissolve. You could ask questions such as the following to bring attention to the variables in this test.

- How many cups of water do we need?
- Should the cups all have the same amount of water?
- What else about the water should be the same? (same temperature)
- Should you swirl the cups in the same way and for the same length of time?

3. Have students dissolve the crystals in water.

The amount of water used in the following procedure is specific and should be used because it gives clear results. It is helpful if you lead the class so that all groups pour their crystal samples into the water at the same time. You could also count off the 20 seconds for each session of swirling so that all groups can compare their results at the same time.

Procedure

1. Place 1 teaspoon of hot tap water in each empty clear plastic cup.

2. Match up each pair of cups so that each cup of crystal is near its corresponding cup of water. At the same time, with the help of your lab partners, pour the weighed amount of crystals into its cup of water.

3. With the help of your lab partners, swirl each cup at the same time and in the same way for about 20 seconds and observe. Swirl again for 20 seconds and observe and then for 20 more seconds and make your final observations.

4. Slowly and carefully pour the solution from each clear plastic cup back into its small empty cup. Try not to let any undissolved crystal go into the small cup. Compare the amount of crystal remaining in each clear plastic cup.

Pass out *Activity sheet 2.3—Solubility test* on page 62, so that students can record their observations. Students also will need to consider the possible identity of the unknown based on this test as they continue to document the process of identifying the unknown on *Activity sheet 2.1, 2.2, 2.3, 2.4—Using physical change to identify an unknown solid* on page 61.

4. Have students discuss their observations.

Ask students questions such as the following:

- What do you think is the identity of the unknown?
- What evidence do you have to support your conclusion?
- If someone in the class had a very different conclusion and had very different observations, what do you think may have led to these differences?

Students should mention possible errors in weighing the crystals, in measuring the amount of water used, or the amount and type of stirring, or accidentally pouring the crystals into the wrong cups.

Expected results: **Results may vary somewhat depending on the temperature of the water. However, Epsom salt and MSG should appear to dissolve more than salt and the unknown.**

Based on their observations, students are most likely to eliminate Epsom salt as the possible identity of the unknown. They might be able to conclude that the unknown is salt, but in some cases may think it could also be MSG. Since there may be some doubt, students will do a recrystallization test with the crystal solutions from this solubility test. The recrystallization test should be done immediately after the solubility test.

Activity 2.4
Recrystallization test

Question to investigate

Can you identify the unknown crystal by the way it looks when it recrystallizes?

Placing a small amount of each solution on black paper and allowing the water to evaporate will cause recrystallization to occur faster than in the method just described. Some black construction paper works better than others. Before trying a particular type of black paper with your students, follow the procedure below to see how well the crystals show up on your paper.

If no black paper works well, you can make your own by photocopying *Testing sheet 2.4—Recrystallization test* on page 63 and using a black permanent marker to color in the circles.

1. Have students apply the solutions to black paper.

Tell students that they will apply a small amount of each solution to black paper so that the water will evaporate and the crystals will form.

Procedure

1. Use a pencil to label a piece of black construction paper as shown so that there is an area on the paper for each of the 4 solutions.
2. Dip a cotton swab into one of the solutions. Apply the solution in a circular motion to its labeled area to make a wet spot about the size of a quarter. Repeat two more times on this same area so that the spot is very wet.

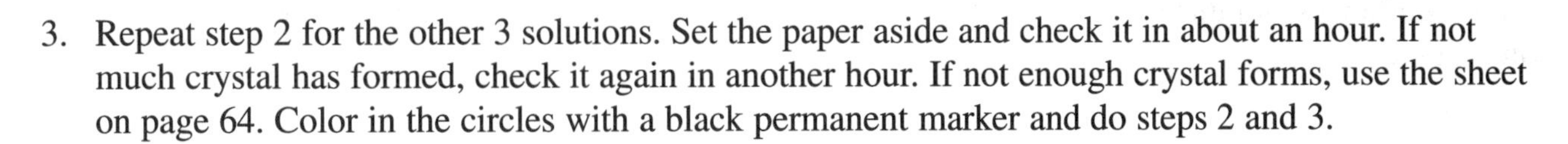

3. Repeat step 2 for the other 3 solutions. Set the paper aside and check it in about an hour. If not much crystal has formed, check it again in another hour. If not enough crystal forms, use the sheet on page 64. Color in the circles with a black permanent marker and do steps 2 and 3.
4. Compare the unknown to the other crystals.

Have students complete *Activity sheet 2.1, 2.2, 2.3, 2.4—Using physical change to identify an unknown solid.* Students should be able to correctly identify the unknown based on this test.

2. Have students discuss their observations.

Students can use a magnifying glass to better see details of the crystals. Ask students what they think is the identity of the unknown. Discuss whether or not they feel they have enough information to be sure.

Expected results: **Salt and the unknown look very similar. MSG and Epsom salt look different from each other and different from salt and the unknown.**

Optional: The 24-hour method

In this activity, students will allow the solutions made in the solubility test to recrystallize. This will take about 24 hours. The crystals that form appear different enough so that students will be able to positively identify the unknown.

1. Have students prepare their clear plastic cups for the recrystallization test.

Procedure

1. Rinse each clear plastic cup with water to remove any remaining crystal. Dry each with a paper towel.

2. Carefully pour the solution from each small cup into its corresponding clear plastic cup.

3. Allow the solutions to sit overnight.

2. The next day, have students observe the crystals.

You may want to have students use a magnifying glass so that they can better see details of the crystals. Have students observe the crystals from the top and bottom of the cup and describe what they see in each cup.

Expected results: **Salt and the unknown look very similar. MSG and Epsom salt look different from each other and different from salt and the unknown.** Have students complete *Activity Sheet 2.1, 2.2, 2.3, 2.4—Using physical change to identify an unknown solid.* Students should be able to correctly identify the unknown based on this test.

When students have completed the activity sheet, ask them questions like the following:

- What do you think is the identity of the unknown?
- Do you have enough information to be sure?

Students should be able to determine that the identity of the unknown is salt. Tell students that the unknown is coarse kosher salt. It is chemically the same as regular salt, but the process for making each is different and that is why they look different.

Name:______________________________

Activity sheet 2.1, 2.2, 2.3, 2.4

Using physical change to identify an unknown solid

Type of test	Circle the possible identity of the unknown. You may circle more than one.	Use evidence from your experiment to explain why the crys-tal(s) you circled could be the unknown.	Use evidence from your experiment to explain why the crystal(s) you did not circle is/are probably not the unknown.
Appearance	Salt Epsom salt MSG		
Crushing	Salt Epsom salt MSG		
Solubility	Salt Epsom salt MSG		
Recrystallization	Salt Epsom salt MSG		

Name:____________________________

Demonstration sheet 2b

Measuring equal amounts of crystals for the solubility test

Even though the *amount* of cereal in each cup is the same, the amount looks different. How do you know that the amount of cereal in each cup is actually the same?

____

____

____

How could accidentally using different amounts of substances make the solubility test unfair?

____

____

____

Name:____________________________

Activity sheet 2.3

Solubility test

Rank the solubility of the following crystals with **1** being the most soluble and **4** being the least soluble.

Salt	MSG	Epsom salt	Unknown
______	______	______	______

Based on this test, what do you think might be the identity of the unknown?

____

Testing sheet 2.4

Recrystallization test

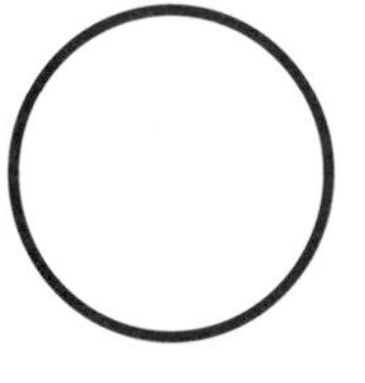

Salt

MSG

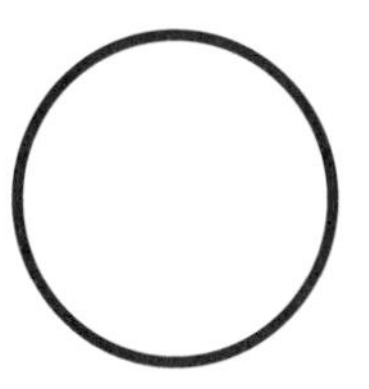

Epsom salt

Unknown

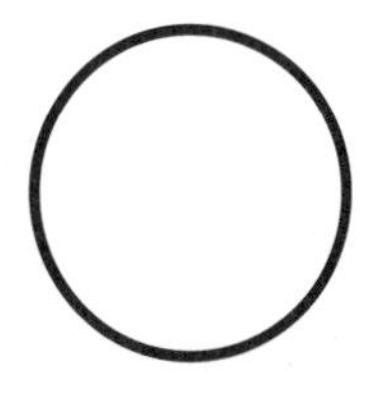

Salt

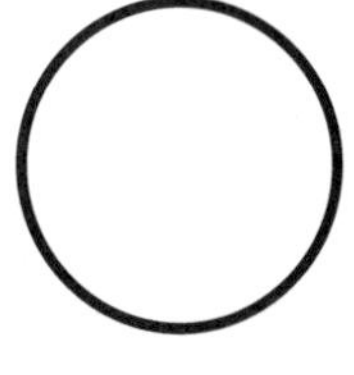

MSG

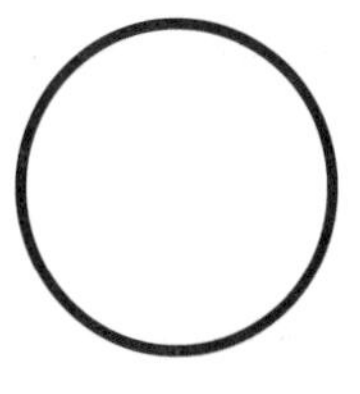

Epsom salt

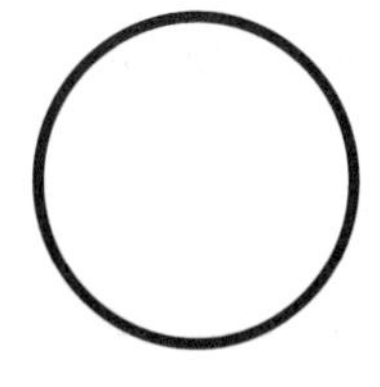

Unknown

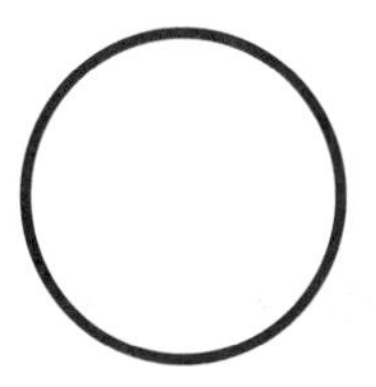

Salt

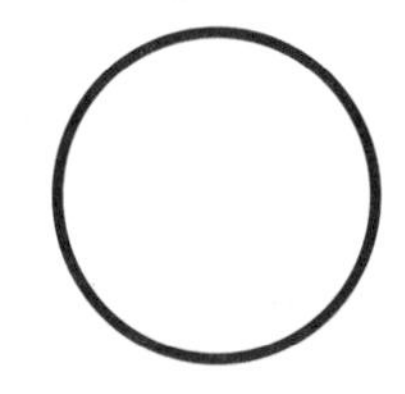

MSG

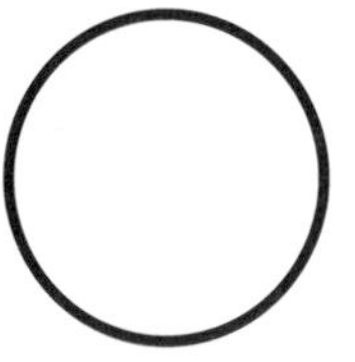

Epsom salt

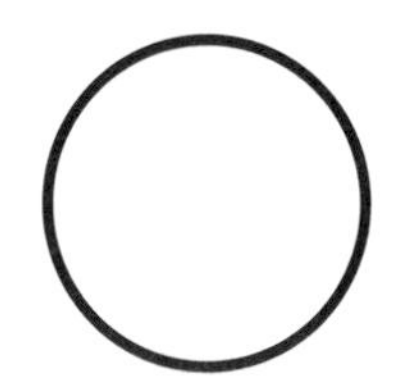

Unknown

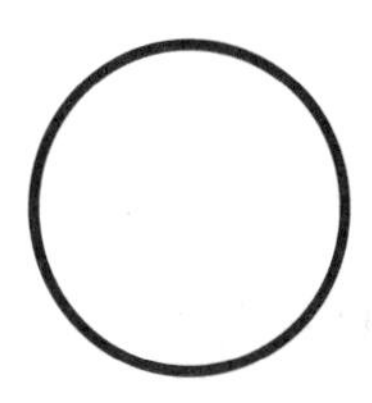

Salt

MSG

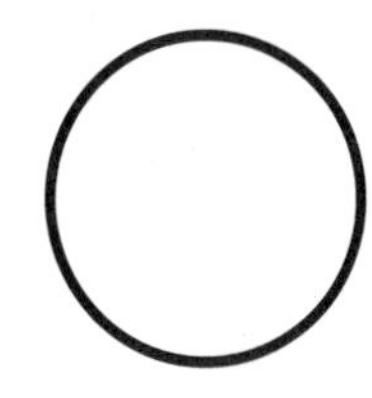

Epsom salt

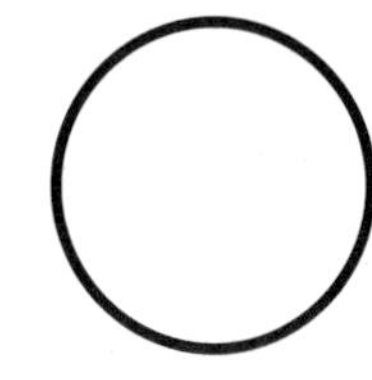

Unknown

Name:______________________________

Assessment rubric

Investigation 2—Physical properties and physical change in solids

To earn a "B," a student must receive a "Very good" in each category.

	Very good	Satisfactory	Needs improvement
Activity sheet 2.1, 2.2, 2.3, and 2.4	________	________	________
Demonstration sheet 2b	________	________	________
Activity sheet 2.3	________	________	________
Investigative behaviors	________	________	________

Activity sheet 2.1, 2.2, 2.3, and 2.4
Using physical change to identify an unknown solid

Circles all possible identities for the unknown after each test
Writes explanations based on evidence from each test
Uses evidence to identify the unknown
Uses evidence to explain why other crystals cannot be the unknown

Demonstration sheet 2b
Measuring equal amounts of crystals for the solubility test

Records actual investigations
Questions are well-written
Uses clear drawings of experimental setup
Answers questions based on observations

Activity sheet 2.3
Solubility test

Correctly ranks solubility of crystals
Reasonably identifies crystals that may be the unknown
Recognizes why weight is better than volume to measure equal amounts of crystals

Investigative behaviors

Participates in design of tests
Conducts fair tests
Works cooperatively with group
Uses evidence to formulate explanations

To earn an "A," a student must also exhibit some of the following qualities throughout this investigation.

Shares detailed observations
Writes outstanding explanations
Participates well in class discussions
Participates well in group work
Uses scientific thinking

Investigation 3.
Physical properties and physical change in liquids

Summary

Students will first conduct tests on four different clear colorless liquids. They will either design their own tests or be guided to design tests that include the appearance of a drop of each liquid on wax paper, an absorption test, and an evaporation test. Students will use at least two of these tests to identify an unknown liquid. The liquids will then be colored with food coloring so that students can combine each colored liquid with a drop of colored water. The characteristic way each liquid combines with water can be used to identify all four unlabeled liquids. The activities will emphasize the characteristic properties of liquids, identifying and controlling variables, making observations, and analyzing results to answer a question.

Objective

Through testing different liquids, students will develop an understanding of the meaning of "characteristic properties" of substances. As students help design valid scientific tests, they will identify possible variables and suggest ways to control them.

Assessment

The assessment rubric *Physical properties and physical change in liquids* on page 85 is included so that you can assess and document student progress throughout the investigation. The abilities and understandings demonstrated by students and recorded on the rubric include controlling variables as students help design and conduct fair tests and recording detailed observations that they will rely on as they identify the unlabeled liquids. Investigative behaviors observed as students design and conduct fair tests and then use the results from these tests to efficiently test the unknowns are also recorded on the rubric.

Two activities in this investigation can also serve as performance assessments. This means that in order for students to successfully complete these activities they must demonstrate mastery of certain concepts. By observing students as they conduct *Activity 3.1—Developing tests to distinguish between similar-looking liquids* and *Activity 3.3—Using the combining test to identify the unlabeled liquids*, you can informally evaluate to what extent students have mastered the abilities, understandings, and investigative behaviors developed in the investigation.

Relevant *National Science Education Standards*

Physical Science

K–4

Properties of Objects and Materials

Objects have many observable properties.

5–8

Properties and Changes of Properties in Matter

A substance has characteristic properties.

Science as Inquiry

K–4

Abilities Necessary to do Scientific Inquiry

Ask a question about objects.
Plan and conduct a simple investigation.
Employ simple equipment and tools to gather and extend the senses.
Use data to construct a reasonable explanation.
Communicate investigations and explanations.

Understandings about Scientific Inquiry

Scientific investigations involve asking and answering a question.
Types of investigations include describing objects…and doing a fair test.
Good explanations are based on evidence from investigations.

5–8

Abilities Necessary to do Scientific Inquiry

Identify questions that can be answered through scientific investigations.
Design and conduct a scientific investigation.
Use appropriate tools and techniques to gather, analyze, and interpret data.
Develop descriptions, explanations, predictions, and models using evidence.
Think critically and logically to make the relationships between evidence and explanations.
Communicate scientific procedures and explanations.

Understandings about Scientific Inquiry

Different kinds of questions suggest different kinds of scientific investigations.
Scientific explanations emphasize evidence and have logically consistent arguments.
Scientific investigations sometimes result in new ideas and phenomena for study or generate new procedures for an investigation. These can lead to new investigations.

How this investigation relates to the *Standards*

This investigation is designed to achieve two related goals of the *Standards*. One is for students to discover that substances have characteristic properties and undergo characteristic physical change. The other goal is for students to begin to develop the skills of designing and conducting fair tests to discover these characteristics. Since these physical properties are particular to each liquid, they can be used to help identify the unknown liquid. Students will discover that in order to compare the properties of different liquids, each liquid needs to be tested in the same way. These ideas concerning identifying and controlling variables are central to any scientific investigation.

Materials chart

3.1 Developing tests to distinguish between similar-looking liquids

3.2 Using color to see how liquids combine

3.3 Using the combining test to identify the unlabeled liquids

Each group will need

	Activities 3.1	3.2	3.3
tap water in cup	•		
alcohol solution in cup	•		
detergent solution in cup	•		
salt water in cup	•		
unknown in cup	•		
4 yellow solutions in cups labeled A, B, C, D			•
blue tap water in cup			•
labeled droppers or straws	•	•	
4 droppers/straws labeled A, B, C, D			•
wax paper	•		
crayons or colored pencils		•	
toothpicks		•	•
brown paper towel or brown coffee filters	•		
absorbent paper towels		•	•
laminated chart		•	•
newspaper/newsprint	•		
colored construction paper	•		

Notes about the materials

The same solutions are used in all three activities in this investigation. However, in activity 3.1 they are colorless. In activities 3.2 and 3.3, they are colored yellow so that students can observe their movement as they combine with tap water, which will be colored blue.

The charts used in activities 3.2 and 3.3 can be found at the end of this investigation on pages 83 and 84. Information about how to prepare these charts is included on page 70.

If using newspaper for activity 3.1, cut strips at least 5 centimeters wide and about 15 centimeters long. Be sure students place drops of the 4 different liquids on similar areas of the paper such as a white unprinted area or an area with regular text. Students should not place the drops on areas that are heavily printed such as pictures in advertisements because the drops will not absorb.

Teacher preparation

The following recipes make ¼ cup of each solution. This is enough for one class to do activity 3.1. Since this set of activities will take more than one day, you may want to make an extra amount of each solution. If you will be able to cover the solutions, you can double the recipes to make the solutions for your class to do both activities 3.2 and 3.3. If not, make fresh solutions for activities 3.2 and 3.3.

Water—Use regular tap water.
Alcohol—Use 70% isopropyl alcohol. This is a common household strength.
Detergent—Add 2 teaspoons colorless liquid hand soap or detergent to ¼ cup tap water.
Salt water—Add 1 tablespoon salt to ¼ cup hot tap water.
Unknown—Any of the solutions may be used. Water can be a challenging unknown.

Note: When using isopropyl alcohol, read and follow all warnings on the label. Be sure students are wearing properly fitting goggles.

Once the solutions are mixed, label and distribute them as follows.

Activity 3.1—Place about 1 teaspoon of each liquid into its labeled cup for each group.

Activity 3.2—Add 2 drops of yellow food coloring to each solution. Then, place about 1 teaspoon of each yellow solution into its labeled cup for each group. Add 4 drops of blue food coloring to ½ cup of tap water. Then, place about 2 teaspoons of this blue water into a cup for each group.

Activity 3.3—Decide which yellow solutions will be A, B, C, and D and write down your choice. Be sure to keep the identity of each solution a secret until students have completed this activity. The following is an example.

A = salt water
B = detergent
C = alcohol
D = water

Place about 1 teaspoon of each yellow solution in its corresponding cup for each group. Add 4 drops of blue food coloring to ½ cup of tap water. Then, place about 2 teaspoons of this blue water into a cup for each group.

Activity sheets

Copy the following activity sheets for this investigation and distribute them as specified in the activities.

Activity sheet 3.1 page 81

Developing tests to distinguish between similar-looking liquids

Copy one per student.

This activity sheet should be given to students after they have conducted their tests on the four liquids. Pass this activity sheet out along with the unknown liquid. As students conduct two of their tests to identify the unknown liquid, they will record information about their testing method on this activity sheet.

Testing sheets 3.2 and 3.3 page 82

Using color to see how liquids combine

Using the combining test to identify the unlabeled liquids

Copy one per group

For best results cut along the dotted lines and laminate the two charts. If you do not have access to a laminating machine, place each chart in a sandwich-size zip-closing plastic bag, seal it, and tape it to the desktop. Or, place a transparency on top of each chart and tape it to the desktop.

Activity sheet 3.2 page 83

Using color to see how liquids combine

Copy one per student.

Students will use this activity sheet to record their observations as they combine each yellow liquid with blue water. They will refer to this activity sheet as they attempt to identify the unknown liquids in *Activity 3.3—Using the combining test to identify the unlabeled liquids.*

Activity sheet 3.3 page 84

Using the combining test to identify the unlabeled liquids

Copy one per student.

On this activity sheet, students will write the identity of each of the four unlabeled liquids. They will also describe the evidence that led them to reach these conclusions.

Science background information

The physical properties of a substance are its characteristics that can be described without changing the substance's identity. For liquids, some physical properties are color, density, and surface tension. Some other physical properties of liquids that may not be as easy to observe and measure are viscosity and how easily they dissolve a given solid.

A concept related to physical properties is physical change. A physical change describes a type of change in a substance that does not change the fundamental identity of the substance. A common physical change for a liquid is changing from one state to another through boiling or freezing. Some others include absorbing into another material, evaporating, or mixing with another liquid to become part of a solution. In all these cases, the liquid changes its form but not its identity. The physical properties of a liquid and the way it undergoes physical change are characteristic properties and can be used to distinguish it from other liquids.

Activity 3.1—Developing tests to distinguish between similar-looking liquids

Liquids on wax paper test

The different appearance of the liquids on wax paper has to do with the liquids themselves as well as the wax paper surface they are placed on. The particles that the different liquids are composed of give each its own characteristic properties. But the way they look when placed on wax paper has to do with their own particular characteristics *and* how they interact with the wax. Since water is a polar molecule with areas of positive and negative charge, water molecules tend to stick to, spread out on, and dissolve other polar substances. But wax is non-polar. Its molecules do not have positively and negatively charged areas. The fresh water and salt water bead up on wax paper because water tends to stick to itself and not to the wax. Because of its polar nature, water molecules tend to associate with each other in a cohesive way contributing to a phenomenon called *surface tension.* These factors cause fresh water and salt water to bead up on the wax paper. Detergent molecules and alcohol molecules each have an area of the molecule that is polar and an area that is non-polar. Neither liquid has the strong surface tension of water, so neither beads up the way water and salt water do and both are more attracted to the wax than water and salt water are.

Absorption test

The degree to which a liquid absorbs into another substance, in this case, a paper towel, has to do with the properties of the liquid and the properties of the paper towel. One aspect of absorbing has to do with the physical properties of the paper. The size and number of pores in the paper make a big difference. In addition to the physical features of the paper, absorbency depends upon the liquid's ability to move through the pores by capillary action. Capillary action depends upon the liquid's ability to adhere to the material and to itself so that it moves up and into the small spaces of the material. On the molecular level, a liquid's ability to move up into the pore spaces of a material will depend upon the attraction of the liquid for the material (adhesion) and on the attraction of the molecules of the liquid for each other (cohesion). Based on these characteristics, the different liquids absorb into the paper towel to different extents.

Evaporation test

Evaporation of a liquid occurs when particles of the liquid gain enough energy to break away from the rest of the liquid and move into the surrounding air. The rate of evaporation depends on the amount of attraction the particles of the liquid have for each other. When the liquids have absorbed into another material, in this case, paper towel, the attraction between the liquid and the molecules that make up the paper towel will also affect the rate of evaporation. The combination of these factors determines the different rates of evaporation for each liquid.

Activity 3.2—Using color to see how liquids combine, and Activity 3.3—Using the combining test to identify the unlabeled liquids

The yellow liquids each combine with blue water differently because of the way the molecules that make up the liquids interact with water molecules.

Water/water

When the blue water and yellow water come into contact, the random motion of the water molecules in both samples causes some mixing, resulting in some green but usually leaving areas of blue and yellow remaining. If given enough time, this movement of water molecules will eventually result in complete mixing and a green color throughout.

Water/salt water

The fresh water and salt water combine more quickly and result in a green color throughout the combined solutions. One explanation is that when the salt dissolves in water, it breaks down or dissociates into positively charged sodium ions and negatively charged chloride ions. The polar water molecules and these charged particles are very much attracted to one another and mix quickly.

Water/alcohol

The isopropyl alcohol is somewhat different. It has an end that is polar and an end that is somewhat non-polar. The end that is polar readily associates with water, and the other end does not. This could account for its different appearance when it interacts with water.

Water/detergent

The different look of the combination of water and detergent could be caused partially by the physical thickness or viscosity of the detergent. It could also result from the charges on the detergent molecules and how they interact with water, detergent's ability to break water's surface tension, or a combination of these or other factors.

Activity 3.1
Developing tests to distinguish between similar-looking liquids

Question to investigate

How can you use simple materials to design tests to help you tell the difference between similar-looking liquids?

This investigation has three activities. In the first activity, students will develop tests to distinguish between four liquids and then use these tests to identify an unknown liquid. Activity 3.1 has two different options for developing the tests. The first option could be characterized as more "open" inquiry, while the second option uses a more "guided" approach.

Option A: More open inquiry

Students will have the opportunity to select from various materials to develop their own tests to tell the difference between four different clear colorless liquids. They will then use these tests to identify an unknown liquid. This unknown is one of the four labeled liquids that students have been testing.

1. Introduce students to the four clear colorless liquids.

Show students samples of each of the four clear colorless liquids in labeled clear plastic cups. Tell students that scientists sometimes have substances like clear liquids that look the same, but they need to be able to tell the difference between them. Tell students that in this activity they will develop tests to try to tell the difference between these similar-looking liquids.

2. Have students consider tests they could try to help them tell the difference between the four liquids.

Show students materials such as wax paper, newspaper, construction paper, brown coffee filters or paper towels, and droppers. Ask students how they could use the liquids and these materials to design tests to help see differences between the liquids. Tell students that after they develop their tests, you will give them an unknown liquid that is one of the four liquids, and that they will use their tests to identify it.

3. In groups, have students develop their own methods to test the liquids and then try these tests.

Talk about ways to prevent contamination of the liquids in the cups. Students should be careful not to put the same dropper in more than one liquid. To avoid mixing up the droppers, you could label them with masking tape or write directly on them with a permanent marker. Students should be sure to use the correct dropper with each solution. To prevent spills, tape the cups to the table so that the cup and dropper do not fall over.

As you observe and talk with groups about their plans, ask them what they are doing to make their tests as fair as possible. Students should plan to use the same amount of each of the four liquids, applied to the same surface, in the same way, at the same time, etc. Some tests that students may develop could include:

- Placing a drop of each liquid on wax paper,

  Drops will either bead up or spread out.
- Tilting the wax paper so that the drops move,

  Drops will run down and leave different-looking marks.
- Placing a drop of each liquid on a brown paper towel, newspaper, brown coffee filter, or piece of construction paper,

  The drops will make different marks as they absorb into the paper.
- Allowing drops to evaporate from a paper towel or piece of construction paper,

  Drops will evaporate at different rates.

Students may develop other tests that work well. The goal is for them to develop at least two different tests that show a distinct enough difference between the liquids so that they can use these tests to identify the unknown liquid.

Teacher note: Based on the way the liquids bead up or spread out on a surface like wax paper, students can easily divide the liquids in two groups—salt water and fresh water, and rubbing alcohol and detergent solution. Salt water and fresh water bead up and alcohol and detergent spread out. Since students will need to look for characteristic differences between all the liquids, they will need to develop another test to distinguish between salt water and fresh water and between rubbing alcohol and detergent solution. A surface such as newspaper or newsprint or construction paper can help students distinguish between these pairs of liquids. They can observe some differences in the way each liquid absorbs into the paper and differences in the way each liquid evaporates from the paper. As you interact with groups as they design and conduct their tests, encourage students to conduct fair tests that control variables. For instance, students should place one drop of each liquid on the same surface at the same time and observe each liquid very closely. Point out that there are subtle differences in the way *each* of the liquids behave on each surface. When students conduct an absorption test, you may need to suggest that this test can also be used as an evaporation test if students observe it for a long enough period.

4. Have students share their testing methods and observations.

Students should realize that there are a variety of different ways to tell the difference between the liquids. Point out that although the methods they developed are different, they should all have one thing in common: They should control the variables as much as possible to make a fair test.

5. Give groups the unknown liquid to identify.

Pass out *Activity sheet 3.1—Developing tests to distinguish between similar-looking liquids* to each student as you give each group their unknown liquid. Direct students to pick two tests that they think will allow them to identify the unknown. In each of the two tests, have students use all four known liquids, along with the unknown liquid.

Teacher note: If you selected fresh water for the unknown, distinguishing it from salt water can be a challenge. Fresh water absorbs into brown paper towel, brown coffee filter, newsprint, and especially construction paper a bit faster than salt water. In an evaporation test, fresh water will evaporate a little bit more quickly than salt water.

6. Ask students for the identity of the unknown.

Ask students what they think is the identity of the unknown and explain what evidence led them to that conclusion.

Option B: More guided inquiry

If students need more direction than is offered in Option A, you may try presenting this more guided option. Here, students will conduct two tests to learn about the physical properties of the four liquids. They will then use these tests to identify an unknown liquid.

Liquids on wax paper test

Students will place drops of four different clear colorless liquids on a piece of wax paper. They will observe what each liquid looks like and how it moves on the wax paper.

1. Have students carefully look at four clear colorless liquids.

Ask students how they could tell two similar-looking liquids apart. For example, if you gave them a sample of fresh water and a sample of salt water and no tasting was allowed, what could they do to tell them apart? Students may suggest evaporating the liquids, seeing if they bead up on a surface, seeing if they absorb differently, or other ideas.

Have students look at the *fresh water*, *salt water*, *alcohol*, and *detergent solution* in their cups. Ask them if they can tell them apart, just by looking. They should not be able to.

2. Have students place a small amount of each liquid on a piece of wax paper.

Talk about ways to prevent contamination of the liquids in the cups. Students should be careful not to put the same dropper in more than one liquid. To avoid mixing up the droppers, you could label them with masking tape or write directly on them with a permanent marker. Students should be sure to use the correct dropper with each solution. To prevent spills, tape the cups to the table so that the cup and dropper do not fall over.

Procedure

1. Use a pencil to label a piece of wax paper with the names of each of the solutions. Use a dropper to place a small amount of each liquid near its label.

2. Tilt the wax paper so that the drops slide down the paper a bit. Lay the wax paper flat again and wipe the drops off with a paper towel.

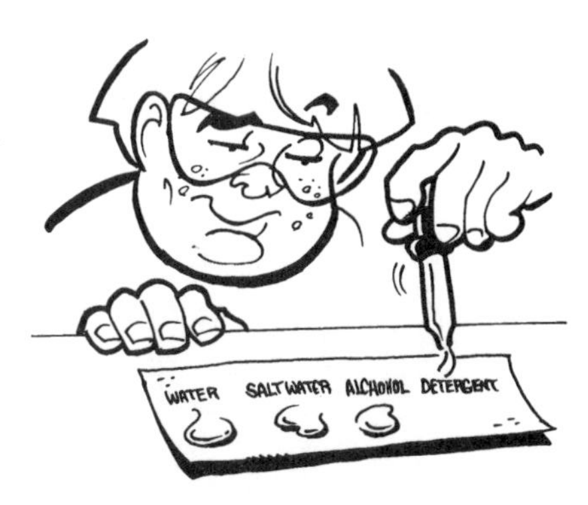

3. Have students report their observations.

Ask students, if they noticed any similarities and differences among the drops when they…

- were first placed on the wax paper,
- slid down the wax paper, and
- were wiped off the wax paper.

Expected results: **The water and salt water will bead up on the wax paper, while the detergent and alcohol will spread out and be flatter. When the drops are wiped off the paper, the water and salt water may leave a circular mark and the detergent will leave a streak where it traveled down the wax paper.**

Absorption/Evaporation test

Students will compare the way that each solution absorbs into a brown paper towel. They will be looking at the size, shape, color, and speed of spreading of the mark made by each liquid. Students will see differences, some more obvious than others, between the absorbing liquids. They will discover that after a certain period of time the absorption test can also become an evaporation test. Students will observe that the liquids evaporate from the paper towel at different rates. Students should conclude that both the absorption and evaporation tests could be used to help identify the unknown liquid.

1. Have students apply the liquids to the paper towel.

Have a class discussion to help students design a simple absorption test. Discuss with students how they will control the variables to make the absorption test as fair as possible. You should emphasize applying the same amount of each liquid, on the same surface, in the same way, at the same time, and for the same length of time. The following procedure is an example of a procedure that you and your students might design. Feel free to follow any procedure you and the students develop. You may choose to place the liquids on a surface such as a brown paper towel, brown coffee filter, newspaper, or construction paper. Test these ahead of time and choose the surface that you feel shows the most obvious difference between the liquids.

Procedure

1. Use a pencil to label 4 areas on a piece of brown paper towel as shown.

2. Pick up a small amount of each liquid in its dropper.

3. At exactly the same time, squeeze one drop of each liquid in its labeled area. Watch the liquids absorb for about one minute.

2. Ask students to share their observations.

Ask students what similarities and differences they notice in the drops as they absorb into the paper towel.

Expected results: **The fresh water makes the largest mark on the paper towel and the detergent makes the smallest mark. The fresh water seems to absorb into the paper towel a little faster than the salt water, which stays beaded up on the paper towel a little longer than the water. After they absorb, the fresh water and salt water appear similar, but the fresh water seems to wet a larger area. The detergent and alcohol leave smaller marks on the paper towel. The alcohol mark has a smooth edge, while the detergent has a more irregular edge. Also, the detergent mark appears darker.**

Remind students that in the wax paper test, they noticed that fresh water and salt water were similar and that alcohol and detergent were similar. Ask students how this absorption test helps them distinguish between these sets of liquids a little more.

3. Have students continue observing the drops as an evaporation test.

Students may have noticed that one or more of the marks made by the liquids are becoming lighter in color. Ask students what they think might be happening. Remind students of other examples of things that are wet becoming dry and tell them that this process is called *evaporation*. Ask students what differences they notice in the marks as the liquids evaporate. Then ask whether this *evaporation test* is a fair test and what makes it fair. Students should mention that they placed the same amount of liquid on the same surface at the same time. The liquids were also exposed to the same air and temperature for the same length of time. Since these variables were controlled, this evaporation test is a fair test. After agreeing that the evaporation test can also help students tell the difference between the liquids, have them share their observations for the evaporation test.

Expected results: **The rubbing alcohol evaporates first. Although the fresh water and salt water are similar, the fresh water evaporates before the salt water. The detergent solution evaporates the slowest.**

Identify the unknown liquid

Give each group an unknown liquid that is the same as one of the four liquids they have been testing. Water works well as the unknown. Also pass out *Activity sheet 3.1—Developing tests to distinguish between similar-looking liquids* to each student. Based on students' observations and experience conducting the tests, they should be able to identify the unknown liquid.

1. Have groups conduct tests to identify the unknown liquid.

Students should conduct the tests as they did before, but include the unknown liquid as one of the samples. By observing the way the unknown liquid looks on wax paper, absorbs into a paper towel, or evaporates, students should be able to identify the unknown.

2. Ask students for the identity of the unknown.

Ask students what they think is the identity of the unknown and have them explain what evidence led them to that conclusion.

Activity 3.2
Using color to see how liquids combine

Question to investigate

Can you tell a difference between liquids by the way they look when they combine with water?

In the next two activities, students will use a different method to observe another characteristic property of water, saltwater, alcohol, and detergent solution. The liquids will be colored yellow so that students can observe the different ways they combine with water that has been colored blue. After carefully observing the combining liquids in this activity, students will be able to identify these same yellow liquids, but unlabeled, in Activity 3.3.

1. Have students combine a small amount of each of the yellow liquids with blue water.

As you explain the following procedure, discuss with students the importance of using each pipet only for one liquid to be sure that liquids are not accidentally mixed in the cups. Students should also use a clean toothpick as they combine each pair of liquids. Point out that the test would be most fair if they pulled the blue liquid to the yellow liquid each time.

Students should use *Testing sheet 3.2—Using color to see how liquids combine* to conduct this test. They should record their observations on *Activity sheet 3.2—Using color to see how liquids combine* as each liquid combines with water. This completed activity sheet will be used in Activity 3.3 as students attempt to identify the four unlabeled yellow liquids.

Procedure

1. Add drops of each liquid to its labeled circle to completely fill each circle on the chart. Depending on your dropper, you may need to add about 5 drops or more.

2. Then, use a toothpick to pull the blue water toward the yellow water. It may take a few tries to get them to join. As soon as the 2 drops meet, lift the toothpick away and discard it. Watch the 2 drops combine on their own. *Do not stir.*

3. Record your observations.

Teacher note: Students should use yellow and blue pencils or crayons to draw what the blue and yellow water looked like as they were combining. They should also write a descriptive caption for each drawing that gives information that their drawing can't, like "the ends stayed yellow and blue for a while with a green area in the middle."

4. When the drawings and captions are complete for the first pair, combine the second pair of drops and record your observations.

5. Continue testing the remaining pairs in this manner.

2. Have students discuss their observations.

Ask students to describe what happened as each pair of liquids combined.

Expected results:

- **water + water**

Colors do not combine completely. There will be a region that turns green in the middle, while the ends of the merged drops remain the original yellow and blue.

- **water + salt water**

Colors combine almost immediately as evidenced by the quick change to green throughout.

- **water + alcohol**

The yellow and blue drops appear to "shake" for a time as they combine.

- **water + detergent**

Colors combine at a medium rate resulting in a more spread-out area of light green.

Ask students if they think they could use this test to identify the four yellow liquids even if they were unlabeled.

Activity 3.3
Using the combining test to identify the unknown liquids

Question to investigate

How can you use the characteristic way each liquid combines with water to identify unknown liquids?

In this activity, students will use the combining test to identify the same four yellow liquids, but this time they are unlabeled. By testing each liquid with blue water as in Activity 3.2, comparing the results to their recorded observations, and conducting new tests, students will identify the four unknown liquids.

1. Have groups test and identify all four unknown liquids.

Students should use *Testing sheet 3.3—Using the combining test to identify the unlabeled liquids* to conduct this test. They will also need the completed *Activity sheet 3.2—Using color to see how liquids combine*. Pass out *Activity sheet 3.3—Using the combining test to identify the unlabeled liquids* so that students can record the identity of each of the unlabeled liquids as they discover them.

Students may develop and follow their own procedure to identify the unlabeled liquids. The following procedure is one example.

Procedure

1. Use drops of yellow liquid labeled A to fill its circle and then fill the opposite circle with blue water. Combine this pair with a toothpick the way you did in the last activity.

2. To figure out the identity of unknown A, compare the way these two liquids combine to your drawings and captions from the previous activity.

What do you think is the identity of unknown A?

3. When you think you might know, test that liquid with blue water on your other chart and compare it to the way unknown A combines with blue water. Conduct the test as many times as you need to.

4. Repeat steps 1–3 for each of the remaining unknowns. Record the identity of each unknown liquid on your observation sheet and include what made you think that was the identity of the unknown.

2. Have students compare their results and conclusions.

Have students discuss their results and say what they think are the identities of the unknown liquids. Ask students what evidence they used to help them decide the identity of each liquid. Reveal the identities of all four of the unknown liquids.

Name:______________________________

Activity sheet 3.1

Developing tests to distinguish between similar-looking liquids

Your teacher gave you an unknown liquid. This liquid is the same as one of the liquids you have been testing in this investigation. Your job is to use any 2 tests you have conducted to identify the unknown liquid.

Name of test ______________________	Name of test ______________________
Briefly explain your procedure for testing the unknown. ______________________ ______________________ ______________________ ______________________	Briefly explain your procedure for testing the unknown. ______________________ ______________________ ______________________ ______________________
What did you try to keep the same in your procedure to make it a fair test? ______________________ ______________________ ______________________ ______________________	What did you try to keep the same in your procedure to make it a fair test? ______________________ ______________________ ______________________ ______________________
After doing this test, what did you think was the identity of the unknown? ______________________	After doing this test, what did you think was the identity of the unknown? ______________________
What evidence did you observe to make you think this? ______________________ ______________________ ______________________ ______________________	What evidence did you observe to make you think this? ______________________ ______________________ ______________________ ______________________

Testing sheet 3.2

Using color to see how liquids combine

1	2	3	4
Blue	Blue	Blue	Blue
Water	**Water**	**Water**	**Water**
Water	**Salt Water**	**Alcohol**	**Detergent Solution**
Yellow	Yellow	Yellow	Yellow

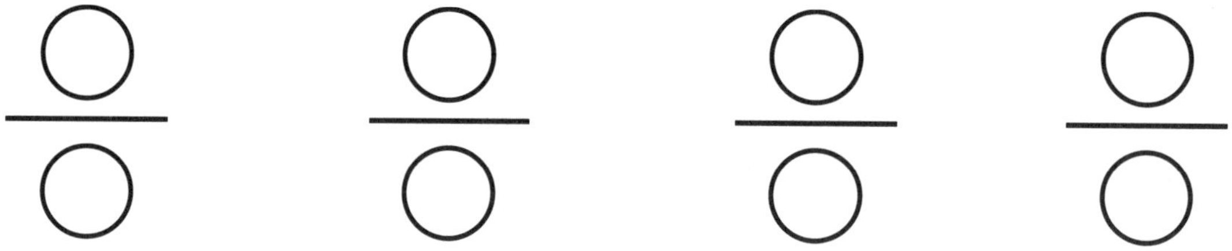

Testing sheet 3.3

Using the combining test to identify the unlabeled liquids

Blue	Blue	Blue	Blue
Water	**Water**	**Water**	**Water**
A	**B**	**C**	**D**
Yellow	Yellow	Yellow	Yellow

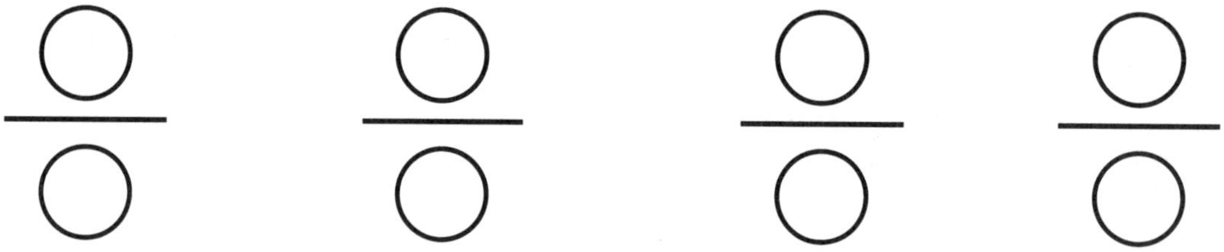

Name:________________________________

Activity sheet 3.2

Using color to see how liquids combine

Draw what the drops looked like when they first joined together.	Write words and phrases that describe your observations
Blue Water ○ Yellow Water ○	
Blue Water ○ Yellow Salt Water ○	
Blue Water ○ Yellow Rubbing Alcohol ○	
Blue Water ○ Yellow Detergent Solution ○	

Name:_______________________

Activity sheet 3.3

Using the combining test to identify the unlabeled liquids

What is the identity of each unknown?		What clues helped you identify the unknown?
A		
B		
C		
D		

Name:______________________________

Assessment rubric

Investigation 3—Physical properties and physical change in liquids

To earn a "B," a student must receive a "Very good" in each category.

Very good | Satisfactory | Needs improvement

Activity sheet 3.1 ________ ________ ________

Developing tests to distinguish between similar-looking liquids

Creates drawings that represent tests well
Writes clear procedures
Discusses controlling variables
Uses evidence to help identify the unknown

Activity sheet 3.2 ________ ________ ________

Using color to see how liquids combine

Records observations with detailed drawings
Uses descriptive words and phrases to supplement drawings

Activity sheet 3.3 ________ ________ ________

Using the combining test to identify the unlabeled liquids

Uses evidence from observations to identify the unknowns
Identifies all unknowns correctly

Investigative behaviors ________ ________ ________

Participates in design of tests
Conducts fair tests
Works cooperatively with group
Uses evidence to formulate explanations
Uses logic to test efficiently

To earn an "A," a student must also exhibit some of the following qualities throughout this investigation.

Shares detailed observations
Writes outstanding procedure
Participates well in class discussions
Participates well in group work
Uses scientific thinking
Consistently exhibits exceptional thought and effort in tasks

Investigation 4. Chemical change

Summary

The way substances react chemically when combined with test liquids is a characteristic property that can be used to distinguish between similar-looking substances. In this investigation, students will design a method for testing five similar-looking substances with four test liquids and observe the reactions. Students will then use the characteristic set of reactions for each substance to help identify an unknown substance.

Objective

Students will realize that the way a substance reacts chemically is a characteristic property of that substance. They will design an organized testing procedure, test the substances in a consistent way, record their observations, and use the characteristic chemical changes to correctly identify an unknown substance.

Assessment

The assessment rubric *Chemical change* on page 102 is included so that you can assess and document student progress throughout the investigation. The abilities and understandings demonstrated by students and recorded on the rubric include the following: Students will consider how to best organize and conduct chemical tests on different powders and record their observations. Students will then be able to use the characteristic chemical changes that may occur to help them identify the unknown powder. Investigative behaviors observed as students help design an organized testing procedure and conduct fair tests are also recorded on the scoring rubric.

The challenge of identifying the unknown powder can serve as a performance assessment. This means that in order for students to successfully identify the unknown, they must demonstrate mastery of the experimental design, use of detailed observations, and the use of evidence to form a conclusion. By observing students as they conduct the activity, you can informally evaluate to what extent students have mastered these abilities.

Relevant *National Science Education Standards*

Physical Science

K–4

Properties of Objects and Materials

Objects have many observable properties including...the ability to react with other substances.

5–8

Properties and Changes of Properties in Matter

A substance has characteristic properties.
Substances react chemically in characteristic ways.

Science as Inquiry

K–4

Abilities Necessary to do Scientific Inquiry

Ask a question about objects.
Plan and conduct a simple investigation.
Employ simple equipment and tools to gather and extend the senses.
Use data to construct a reasonable explanation.
Communicate investigations and explanations.

Understandings about Scientific Inquiry

Scientific investigations involve asking and answering a question.
Scientists use different kinds of investigations depending on the questions they are trying to answer.
Types of investigations include describing objects...and doing a fair test.
Good explanations are based on evidence from investigations.

5–8

Abilities Necessary to do Scientific Inquiry

Identify questions that can be answered through scientific investigations.
Design and conduct a scientific investigation.
Use appropriate tools and techniques to gather, analyze, and interpret data.
Develop descriptions, explanations, predictions, and models using evidence.
Think critically and logically to make the relationships between evidence and explanations.
Communicate scientific procedures and explanations.

Understandings about Scientific Inquiry

Different kinds of questions suggest different kinds of scientific investigations.
Scientific explanations emphasize evidence and have logically consistent arguments.
Scientific investigations sometimes result in new ideas and phenomena for study or generate new procedures for an investigation. These can lead to new investigations.
Think critically and logically to make the relationship between evidence and explanations.

How this investigation relates to the *Standards*

One of the goals of the *Standards* is for students to understand that substances react chemically in characteristic ways. In the following activities, students will design a testing procedure and use their observations of the characteristic chemical reactions of different substances to draw reasonable conclusions as to the identity of the unknown substance.

Materials chart

4.1. Using chemical change to identify an unknown

Each group will need

Each group will need	Activity 4.1
baking powder in cup	•
baking soda in cup	•
cream of tartar in cup	•
tide laundry detergent in cup	•
cornstarch in cup	•
unknown (Baking powder) in cup	•
water in cup	•
dilute iodine solution in cup	•
vinegar in cup	•
red cabbage indicator in cup	•
popsicle sticks	6
labeled droppers or straws	4
blank strip of paper	•
labeled laminated paper strips	6

Notes about the materials

When using tincture of iodine, read and follow all warnings on the label.

Use baking powder as the unknown.

Teacher preparation

Teacher will need for the demonstration

zip-closing plastic bag
red cabbage leaves
3 clear plastic cups
laundry detergent
cream of tartar
popsicle stick or small spoon
lukewarm water

Students will need for the activity

1. Prepare the iodine solution by adding about 1 teaspoon of tincture of iodine to ¼ cup water.

2. Prepare the red cabbage solution by tearing up 1 or 2 leaves of red cabbage. Pre-shredded cabbage will not work. Place the pieces in a zip-closing plastic bag. Add about ¾ cup of room-temperature water. Seal the bag and squish the cabbage for about 3–5 minutes until the water turns a medium to dark blue.

3. Label 6 plastic cups baking powder, baking soda, cornstarch, cream of tartar, laundry detergent, and unknown. Place about ½ teaspoon of each powder into its labeled cup. Baking powder works well as the unknown.

4. Label another 4 cups *water*, *iodine*, *vinegar*, and *red cabbage*. Place about 2 teaspoons of each solution into its labeled cup.

Activity sheets

Copy the following activity sheets for this investigation and distribute them as specified in the activity.

Activity sheet 4.1 — page 99
Using chemical change to identify an unknown

Copy one per student.

Students will use this activity sheet to record their observations during each test. While students can look at the results of how the actual powders reacted with the liquids to help them identify the unknown, this is not necessarily the most reliable method since the appearance of some of the chemical reactions change over time. It is important for students to record their observations and refer to them when trying to identify the unknown.

Testing sheets 4.1 — pages 100–101
Using chemical change to identify an unknown

Copy one of each page for each group.

Cut along the dotted lines and laminate each strip so that students have 6 different testing strips. If you do not have access to a laminating machine, either place the strips in large plastic bags or cover them with overhead transparency film.

Students will test each powder with the four test liquids on these strips.

Science background information

Activity 4.1—Using chemical change to identify an unknown

The white powders were chosen for this activity because they are common, inexpensive, safe, and have some interesting reactions with some common test liquids. Several of the reactions involve an acid or a base or both. In case you forgot the difference between an acid and a base, you can read the following explanation. Again, this explanation is intended as background information for you and not as a means for explaining activity observations to students.

Water normally has a certain percentage of its molecules breaking apart (dissociating) and recombining. Due to the interaction of water molecules, the hydrogen from one water molecule sometimes dissociates from its original water molecule and joins another water molecule, leaving its electron behind. Since the water molecule it left behind now has only one oxygen and one hydrogen and an extra electron, it is now OH^-. The water molecule that now has the extra proton from the hydrogen is now H_3O^+. These charged particles are called *ions*. In a sample of water, there is a very small percentage of H_3O^+ and OH^- ions compared to regular water molecules. If a substance is added to water and causes the concentration of H_3O^+ to *increase*, that substance is an *acid*. If a substance is added to water and causes the concentration of H_3O^+ to *decrease*, that substance is a *base*.

Powders used in the activity:

Baking soda (sodium bicarbonate) is a base.

Baking powder is a combination of baking soda (a base) and calcium acid phosphate (a weak acid) and cornstarch.

Cream of tartar is an acid.

Powdered laundry detergents have many formulations but contain a base.

Cornstarch is the starch from corn. It is not an acid or a base.

The reactions that result in bubbling are due to the combination of an acid and the carbonate in baking soda, baking powder, and laundry detergent. The gas released in each case is carbon dioxide (CO_2).

The dark color produced when iodine is used is caused by iodine's reaction with starch from cornstarch.

The other color changes observed are the result of red cabbage juice (an acid/base indicator) reacting with various acids and bases or combinations of both.

Activity 4.1
Using chemical change to identify an unknown

Question to investigate

How can you tell similar-looking substances apart based on the way they react with different chemicals?

In the demonstration, students will be introduced to the concept that different substances react chemically in characteristic ways. In the activity that follows, students will add four different test solutions to baking powder, baking soda, cream of tartar, detergent, and cornstarch. They will use their observations from these tests to identify an unlabeled powder that is one of the five powders they have tested.

1. Do a demonstration to show students that they can tell laundry detergent and cream of tartar apart by the way they react with red cabbage juice.

Teacher preparation:

- Add about ⅛ teaspoon cream of tartar to 1 clear plastic cup and about ⅛ teaspoon laundry detergent to another.
- Tear 1–2 red cabbage leaves into small pieces and place them in a zip-closing plastic bag. Add about ⅔-cup warm water and carefully seal the bag. Vigorously squish the bag until the water turns a medium to dark blue.
- Open a corner of the zip-closing bag and carefully pour only the cabbage juice into a third clear plastic cup.

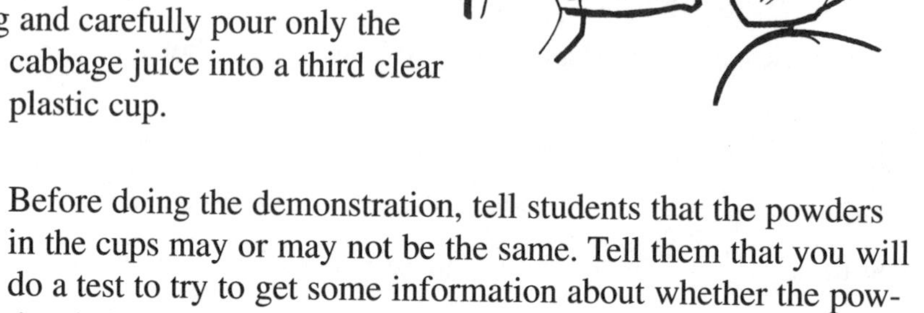

Before doing the demonstration, tell students that the powders in the cups may or may not be the same. Tell them that you will do a test to try to get some information about whether the powders in the cups are the same or different.

Procedure

1. Pour about ⅓ of the cabbage juice into the cup with the cream of tartar and swirl.

2. Pour about ⅓ of the cabbage juice into the cup with the laundry detergent and swirl.

2. Discuss the results of the demonstration.

Ask students if they think that these powders are the same or different. Then ask them what evidence they have for their conclusion.

Expected results: **The cream of tartar will turn the cabbage juice pink, and the laundry detergent will turn it green.** Tell students that although the powders look similar, the fact that they cause the cabbage juice to change to different colors shows that the powders are chemically different.

Tell students that they will use cabbage juice and other test liquids in the following activity to see the characteristic ways they react with, or don't react with, five different similar-looking powders.

3. Have students help decide how they will test the baking powder with the four liquids.

Ask students how they would test the baking powder with each test liquid. They should consider the following questions:

- **Do we need more than one pile of baking powder?**

It would be best to test each liquid on a separate pile of baking powder.

- **How many piles of baking powder should we make?**

Since there are 4 liquids, students should make four piles of baking powder.

- **Do the piles have to be about the same size?**

The size of the piles is not particularly important as long as enough powder is used to see a reaction, if there is one. However, it may be easier for students to fairly compare the results of the unknown to the results from each of the powders if the piles are of similar size.

- **Why is it important to keep track of which piles have been tested with which liquids?**

This is necessary when trying to identify the unknown. If the unknown reacts similarly to one of the powders when a certain solution is added, the unknown may be that powder. Comparing the way each of the piles change when each of the four solutions is added will help to identify the unknown. Labeling will help students know which piles to compare.

- **Should the number of drops placed on each pile be the same?**

The precise number of drops is not particularly important although enough liquid should be added to see the reaction if there is one. However, it will be easier for students to fairly compare the results of the unknown to the results from each of the powders if they use the same number of drops on each pile.

4. Test baking powder with *water*, *iodine*, *vinegar*, and *red cabbage juice*.

Have your students follow their class plan for setting up, labeling, and testing the baking powder. The procedure provided below is one example of a possible plan.

Procedure

1. Use a Popsicle stick to place 4 equal piles of baking powder on the labeled laminated strip. Place a blank strip of paper next to the laminated strip and write the name of each test liquid next to each pile. Label your observation chart the same way.

2. Test each pile of baking powder with 5 drops of each liquid.

5. Ask students to share their results for each test liquid.

As you discuss each group's observations, write the basic results for each test on the board.

Expected results:	**baking powder**
water	bubbling, foamy
vinegar	bubbling, foamy
iodine	black/purple in color, foamy
cabbage	purple then fades to blue, bubbling

6. Plan how to test the other powders so that students can easily compare the results from combining each test liquid to each powder.

Tell students that since the results of the test liquids on the baking powder were the same for each group, then this must be a characteristic set of reactions for baking powder.

Let students know that they will be testing four other powders with the same test liquids and will need to compare the set of reactions for each powder. Ask students what they could do so that it would be easy to see and compare the way each powder reacts with a certain test liquid. Students should realize that powders in the same position on each strip should be tested with the same liquid.

7. Plan how to record the results.

As a class, discuss how students could record the results so that it would be easy to see and compare the way each powder reacted with each particular test liquid. Tell students that it is important to record the results, because the reactions may change over time. Students should realize that a chart, like the one on *Activity sheet 4.1—Using chemical change to identify an unknown* on page 100, is a convenient way to record their results. You may pass out this chart to students or have them create their own chart.

Then, have students write *baking powder* on their observation chart and record the class results for baking powder.

8. Conduct the tests on the remaining powders and record the results.

Procedure

1. Set up the baking soda, cream of tartar, laundry detergent, and cornstarch on laminated strips of paper the way you did with the baking powder.
2. Label the columns on your observation chart with the name of each of these powders.
3. Test each of the powders the way you tested baking powder. Record your observations for each reaction in its corresponding area on your observation chart (*Activity sheet 4.1—Using chemical change to identify an unknown*).

Expected results:	**baking powder**	**baking soda**	**cornstarch**	**detergent**	**cream of tartar**
water	bubbling foamy	no change	no change	no change	no change
vinegar	bubbling foamy	lots of bubbling ends quickly	no change	a little sudsy-looking	no change
iodine	black/purple in color foamy	stays orange	turns black	yellow-ish	stays orange
cabbage	purple then fades to blue bubbling	stays blue	stays blue	turns green	turns pink

9. Test the unknown powder and try to identify it.

Teacher preparation: Place about ½ teaspoon of baking powder in a cup for each group. Label this cup *unknown*.

Give students the unknown powder and tell them that it is one of the five powders they have tested. Students may follow their own testing procedure to identify this unknown powder. However, they should test the unknown powder the same way they tested all the other known powders.

Procedure

1. Place four samples of your group's unknown powder on a separate strip of laminated paper.
2. Test the unknown with each test liquid in the same way you tested the other powders.
3. Compare your observations with the other test strips and with your written observations.

10. Have students report on the identity of the unknown and discuss what evidence led them to their conclusion.

Ask each group to state what they think is the identity of the unknown. Then ask them which observations led them to their conclusion. Ask students if color alone was enough to identify the unknown, or if they needed to use other observations such as bubbling to help identify the unknown.

Name:______________________________

Activity sheet 4.1
Using chemical change to identify an unknown

Observation sheet		Powder being tested					
Test liquid							

Testing sheet 4.1

Using chemical change to identify an unknown

Cream of tartar

Baking soda

Baking powder

Testing sheet 4.1

Using chemical change to identify an unknown

Cornstarch

Detergent

Unknown

Name:______________________________

Assessment rubric

Investigation 4—Chemical change

To earn a "B," a student must receive a "Very good" in each category.

	Very good	Satisfactory	Needs improvement
Activity sheet 4.1 Using chemical change to identify an unknown	________	________	________
Investigative behaviors	________	________	________

Activity sheet 4.1
Using chemical change to identify an unknown

Labels chart correctly
Uses an organized testing and recording method
Uses evidence to help identify the unknown
Writes detailed accurate observations

Investigative behaviors

Participates in design of testing procedure
Participates in design of chart
Conducts fair tests
Works cooperatively with group
Uses evidence to formulate explanations

To earn an "A," a student must also exhibit some of the following qualities throughout this investigation.

Makes and records very detailed observations
Uses logic and evidence to correctly identify the unknown
Participates well in class discussions
Participates well in group work
Uses scientific thinking
Consistently exhibits exceptional thought and effort

Investigation 5. States of matter

Summary

Students will observe warm water changing state through evaporation to become water vapor—a gas. Students will also cool water vapor until it condenses back to liquid water and cool it further until it freezes to become a solid—ice. These changes in water, from a liquid to a gas, gas to a liquid, and liquid to a solid, will show students how temperature affects the states of matter.

Objective

Students will learn that substances can exist in different states—solid, liquid, and gas—and that substances can change from one state to another by heating and cooling.

Assessment

The assessment rubric *States of matter* on page 131 is included so that you can assess and document student progress throughout the investigation. The abilities and understandings demonstrated by students and recorded on the rubric include the following: Students will begin to develop an understanding of evaporation, condensation, melting, and freezing as changes in states of matter. They will realize that changes in temperature can cause changes in state and recognize familiar examples of state changes in everyday life. Investigative behaviors observed as students help design and conduct investigations and develop explanations based on evidence from these experiences are also recorded on the scoring rubric.

Relevant *National Science Education Standards*

Physical Science

K–4

Properties of Objects and Materials

Materials can exist in different states—solid, liquid, and gas. Some common materials, such as water, can be changed from one state to another by heating and cooling.

5–8

Properties and Changes of Properties in Matter

A substance has characteristic properties.

Science as Inquiry

K–4

Abilities Necessary to do Scientific Inquiry

Ask a question about objects.
Plan and conduct a simple investigation.
Employ simple equipment and tools to gather and extend the senses.
Use data to construct a reasonable explanation.
Communicate investigations and explanations.

Understandings about Scientific Inquiry

Scientific investigations involve asking and answering a question.
Scientists use different kinds of investigations depending on the questions they are trying to answer.
Types of investigations include describing objects...and doing a fair test.
Good explanations are based on evidence from investigations.

5–8

Abilities Necessary to do Scientific Inquiry

Identify questions that can be answered through scientific investigations.
Design and conduct a scientific investigation.
Use appropriate tools and techniques to gather, analyze, and interpret data.
Develop descriptions, explanations, predictions, and models using evidence.
Think critically and logically to make the relationships between evidence and explanations.
Communicate scientific procedures and explanations .

Understandings about Scientific Inquiry

Different kinds of questions suggest different kinds of scientific investigations.
Scientific explanations emphasize evidence and have logically consistent arguments.
Scientific investigations sometimes result in new ideas and phenomena for study or generate new procedures for an investigation. These can lead to new investigations.

How this investigation relates to the *Standards*

One goal of the *Standards* is for students to understand that matter can be changed from one state to another by heating and cooling. Through this investigation, students will discover that a substance, such as water, can exist as a solid, liquid, or gas and that heating or cooling a substance can cause it to change from one state to another. These physical changes do not change the identity of the substance but simply change its form from one state of matter to another. Students will also practice identifying and controlling variables as they change the temperature to investigate its effects on changes of state.

Materials chart

5.1 Temperature affects a gas

5.2 Evaporation and condensation

5.3 Cooling water vapor

5.4 Explore student experiences with condensation:

- a. Moisture on the outside of a cold cup
- b. Moisture on the outside of a cold cup—for dry environments
- c. Breathing on cold windows to make them cloudy

5.5 From gas to liquid to solid

Each group will need	Activities 5.1	5. 2	5.3	5.4a	5.4b	5.4c	5.5
½-pint or ½-liter plastic bottle	•						
liquid dish detergent	•						
room-temperature water	•	•		•	•	•	
hot tap water	•	•	•		•		
short clear plastic cups	3	2	2		1		
tall clear plastic cups		2	2	2	1	2	
ice	2 cubes		1 cube	2 cups	2 cubes	1 cup	2 cups
film canisters with caps					2		
quart-size zip-closing bag				•			
paper towel				•	•	•	•
clean empty metal soup can							•
salt							•
plastic teaspoon	•						•
sturdy spoon							•
magnifying glass		•	•				

Teacher preparation

For Activity 5.5, obtain one standard-sized soup can for each group. Remove the label from each can. If you see any sharp or jagged edges around the top, carefully press them down with pliers and cover the rim with duct tape.

Teacher will need for the demonstrations

5.1 Temperature affects a gas
5.4b Explore student experiences with condensation: Moisture on the outside of a cold cup

	Activities	
	5.1	5.4b
½-liter plastic bottle with lid	•	
hot tap water	•	
short clear plastic cup	1	
tall clear plastic cup		1
ice water		•

Activity sheets

Copy the following activity sheets for this investigation and distribute them as specified in the activities.

Activity sheet 5.1 page 127
Temperature affects a gas

Copy one per student.

On this activity sheet, students will identify gas as the substance inside an "empty" bottle. Students will draw a detergent solution bubble expanding and contracting when the gas is warmed and cooled.

Activity sheet 5.2 page 128
Evaporation and condensation

Copy one per student.

Students will show that they understand the terms *evaporation* and *condensation*. They will also write about the results of their experiment.

Activity sheet 5.3 — page 128
Cooling water vapor

Copy one per student.

On this activity sheet, students will record their observations with drawings and then explain how temperature influenced the rate of evaporation and condensation in their experiment.

Activity sheet 5.4 — page 129
Explore student experiences with condensation

Copy one per student.

This activity sheet can be used after students complete any one of the activities in 5.4. Students will consider their own experience with evaporation and condensation, including that from Activity 5.4, as they explain the role of evaporation and condensation in three different hypothetical situations.

Activity sheet 5.5 — page 130
From gas to liquid to solid

Copy one per student.

On this activity sheet, students will record their observations with a labeled drawing and write the possible changes in state that may occur when matter is heated or cooled.

Science background information

Matter exists in three common states or phases: solid, liquid, and gas. Solids, liquids, and gases consist of particles that are in constant motion. In solids, the particles are close to each other and vibrate about fixed positions arranged in an orderly pattern. These properties give solids their definite shape and volume.

When a solid is heated, the motion of the particles it is composed of increases. If enough heat is added, these particles move more freely among each other, and the solid begins to change its state to a liquid. This process is called *melting*. Compared to their movement and position in a solid, the particles in a liquid move more freely and are only slightly further apart.

When a liquid is heated, the motion of the particles increases. When this motion overcomes the attraction the particles have for each other, some particles of the liquid go into the air in the form of a gas. This process is *evaporation*. The rate of evaporation can be increased by increasing the temperature to the boiling point. When essentially all the attractions between the particles is overcome, the substance has become a gas. In this state, the particles move most freely and are very far apart.

Melting and evaporation/boiling require an *input* of energy. The processes of changing from a gas to a liquid and from a liquid to a solid are called *condensation* and *freezing* respectively and result in the *release* of energy.

Activity 5.1—Temperature affects a gas

When the bottle is placed in the hot tap water, the heat adds energy to the air inside the bottle. This increased energy increases the motion of the molecules that make up the air in the bottle. This increased motion causes the molecules to spread out or expand. As the gas expands, some of it escapes from the top of the bottle, which makes the lid move.

When students place the bottle with detergent on it in hot water, energy, in the form of heat, is added to the molecules that make up the air inside the bottle. This added energy increases the molecular motion of the molecules, causing them to spread out or expand. Since there is a detergent solution film over the top of the bottle, the expanding air makes a bubble form on the bottle.

When the bottle is placed in ice water, energy is removed from the air in the bottle. This results in the molecules slowing down, moving closer together, and the gases in the bottle contracting.

Activity 5.2—Evaporation and condensation

The cup covering the hot tap water gets cloudy-looking, and the cup covering the room-temperature water does not. When students check the inside of the cloudy-looking upper cup, it feels wet.

Since the water in one of the lower cups is hot, there is more molecular motion in that sample of water. With more motion, it is more likely that water molecules will have enough energy to break the bonds holding them to the rest of the water molecules and to enter the air inside the upper cup. That is what happens, causing more evaporation from the cup of hot water, creating more water vapor in the upper cup over the hot water.

As the molecules that make up the water vapor in the upper cup contact the inside of the cup and cool, their motion is slowed down and they begin to condense to form liquid water again. So the processes of evaporation and condensation are both occurring in the hot water sample. They are also occurring in the room temperature sample but at a slower rate.

One thing to remember is that temperature is not the only factor that affects the rate of evaporation. Another important factor is the amount of moisture already in the air. Even at higher temperatures, if the air already contains a high concentration of water vapor, the rate of evaporation will not necessarily increase. Air pressure also affects the process of evaporation.

Activity 5.3—Cooling water vapor

The ice cube on the cup cools the area at the top of the cup, which slows the movement of the molecules that make up the water vapor. As the molecules slow down, they tend to move closer together or *condense* and form liquid water again.

Activity 5.4a—Explore student experiences with condensation: Moisture on the outside of a cold cup

This activity explores the common phenomenon of water vapor condensing on the outside of a cold cup. In the previous activities, evaporation from warm water was used as the source for water vapor. In this activity, the source for water vapor is the surrounding air. The cold cup outside the bag is exposed to the water vapor that is naturally in the air. The cold cup inside the bag has a very limited exposure to water vapor in the air. This activity works best if the air is somewhat humid. Because the cup is cold, some heat energy is lost from the water molecules that make up the water vapor around the cup. These molecules slow down and condense on the cup, becoming liquid water. Since the other cold cup in the plastic bag is not exposed to much water vapor from the air, there should be much less evidence of condensation on the outside of this cup.

Activity 5.4b—Explore student experiences with condensation: Moisture on the outside of a cold cup—For dry environments

This activity is similar to 5.4a. It is another way for students to discover that it must be something in the air that causes moisture to collect on the outside of a cold cup. If the air in the classroom is too dry to do Activity 5.4a, you can create a moist-air environment in this activity. The cold film canister placed in the classroom air cools the air around it, but since the amount of water vapor in the air is low, there is little or no observable moisture on the outside of the canister. The cold canister placed under the cup with high water vapor content cools the water vapor, causing it to condense and resulting in more moisture on the outside of the canister.

Activity 5.4c—Explore student experiences with condensation: Breathing on cold windows to make them "cloudy"

In this activity, the outside of both a room-temperature cup and a cold cup are exposed to warm breath. The water vapor in the breath should condense more on the cold cup and form more liquid water than on the room-temperature cup.

Activity 5.5—From gas to liquid to solid

The surface of the can gets so cold that the water vapor in the air around the can condenses to form water, a liquid, and then the liquid water freezes to form ice, a solid. The reason why adding salt to ice makes the ice/water mixture colder than it otherwise would be is somewhat involved, but if you are interested, here is an explanation:

If ice is placed in a well-insulated container, some ice will melt to form liquid water and some liquid water will refreeze to form ice. When there is a balance between the processes of melting and re-freezing, the system is said to be at equilibrium. Melting is a process that requires energy, so as ice melts, it uses up some of the energy inside the container which actually makes the container colder. But freezing is a process that releases energy so as the liquid water re-freezes, it makes the inside of the container warmer. So if the system is at equilibrium, with melting and freezing balancing each other, the temperature in the container will not change. But adding salt interferes with water's ability to refreeze. This upsets the equilibrium so that more ice melts than refreezes. Since the process of melting removes energy from inside the container, it gets colder than it would be at equilibrium.

Activity 5.1
Temperature affects a gas

Question to investigate

Do heating and cooling have an effect on a gas?

In this demonstration and activity, students will be introduced to the idea that a change in temperature effects matter. Although a change in state does not occur when the air inside the bottle is warmed or cooled, students will see that temperature does effect matter. In the activities that follow this one, students will come to understand that sometimes changes in temperature can affect matter so much that a change in state occurs.

1. Do a demonstration to show students that the air in a bottle expands when it is warmed.

Tell students that the air in the bottle is made up of a mixture of gases.

Procedure

1. Add hot tap water to a punch cup until it is about ⅓ full.

2. Use your finger and a little water to moisten the rim of the bottle. Then, place the lid *upside down* on the bottle so that there are no leaks.

3. Carefully push the bottle down into the hot water. After a few seconds, the lid will repeatedly rise and fall with a tapping sound. As you hold the bottle down in the cup of hot water, walk around so that students can both see and hear the movement of the lid.

Teacher note: If you would like to show the demonstration again, you can uncover the opening and let some more air in. Then repeat steps 2 and 3. If the lid does not continue tapping, you may need to "recharge" the bottle by cooling the air inside it. Try rinsing it with a little cold water. Then repeat steps 2 and 3.

2. Discuss with students what may be causing the lid to move.

Ask students what might be causing the lid to go up and down. Tell students that when gases are warmed, they expand. This means that the gas in the bottle needs more room and pushes in all directions, causing the lid to rise off the rim of the bottle.

3. Have students prepare a bubble solution for a related experiment that they will do.

Procedure

1. Mix ½ teaspoon of liquid detergent with about 3 teaspoons of water to make a bubble solution in the cup.

2. Lower the open mouth of the bottle into the bubble solution as shown. Carefully tilt and lift the bottle out so that a film of bubble solution covers the opening of the bottle.

Ask students what they think might happen if they place the bottle in hot water. They should remember their observations from the demonstration and suggest that a bubble will form.

4. Have students place the bottle in hot tap water.

Procedure

1. Add hot tap water to a punch cup until it is about ⅓ full. If the soap film has popped, re-dip the bottle into the detergent solution. Place the bottom of the bottle in the water.

Ask students what they observed when they placed the bottle in hot water. ***Expected results:*** **The bubble film grows into a rounded bubble when the bottle is placed in hot water.** Then ask them what they might do to make the bubble shrink. Students should suggest cooling the air inside the bottle. Ask students why they think cooling the air might cause the bubble to shrink.

5. Have students place the bottle in ice water.

Procedure

1. Put a few pieces of ice in a punch cup and add water until it is about ½ full. Carefully push the bottle down into the cold water.

2. Ask students what they observed when they placed the bottle in cold water.

Expected results: **When the bottle is placed in cold water, the bubble shrinks. It may even go down into the bottle and possibly pop.**

6. Have students record their observations.

When students complete this activity, give them *Activity sheet 5.1—Temperature affects a gas* on page 127.

7. Discuss examples of temperature changing matter.

Ask students if heating and cooling have an effect on a gas. Then ask them if they think temperature could also have an effect on liquids and solids. Students should conclude that temperature does have an effect on a gas. Based on their common everyday experiences, students may be able to conclude that temperature also affects liquids and solids.

Tell students that sometimes temperature can affect a gas, liquid, or solid so much that it could change state. For example, a liquid, like water, could be cooled so much that it changes to a solid (ice) or it could be warmed so much that it changes to a gas (water vapor.) Ask students for more examples in which temperature causes a change in state. Students may suggest examples like water from a wet towel evaporating in the sun until the towel is dry, water vapor in the air cooling to become rain or snow, and snow melting in the sun to become water.

Activity 5.2
Evaporation and condensation

Question to investigate

Does hot water evaporate faster than room-temperature water?

The demonstration and the first activity showed that a change in temperature affects a gas. Temperature also has an effect on a liquid. Warming water increases the rate at which it changes state from *liquid water* to *water vapor*—a gas. This process is called *evaporation*. When enough water vapor collects and cools on a surface, it changes state back to liquid water. This process is called *condensation*.

1. Discuss with students some of their own experiences with evaporation.

Ask students questions like the following to get them thinking about evaporation:

- When you go swimming and get out of the pool, you might decide not to dry off with a towel, but you eventually get dry anyway. Where do you think the water goes?
- The amount of water on your bathing suit is pretty small, but now imagine the amount of water in a big lake. After a long hot dry summer, the level of water in a lake might go down. Where do you think all of that water goes?
- What happens to a pot of water left on a hot stove?

Tell students that in all of these examples the water goes up into the air. This is called *evaporation*.

2. Have students do an activity to compare the evaporation of hot and room-temperature water.

Procedure

1. Fill one punch cup about ⅔ full of hot tap water and fill another about ⅔ full of room-temperature water.

2. Quickly place a tall clear plastic cup upside down over each of the punch cups of water as shown.
3. Watch the cups for about 2 or 3 minutes.
4. Use a magnifying glass to look at the sides and the top of the top cups.
5. Take each tall cup off and feel the inside.

3. Have students record their observations.

When students complete this activity, give them *Activity sheet 5.2—Evaporation and condensation* on page 128.

4. Discuss student observations.

Ask students questions like the following:

- What differences did you notice between the cups over the hot water and the room-temperature water?

  Expected results: **After about 2–3 minutes, the cup over the hot water appears cloudy, while the cup over the room-temperature water does not.**

- What do you think is making the inside of the cup over the hot water cloudy?
- When you felt the inside of the cup over the hot water, it felt wet. How do you think the water got on the inside of the cup?

Explain to students that when water evaporates, it changes from a liquid to a gas, called *water vapor*. In this activity, the hot water evaporated faster than room-temperature water, producing more water vapor in the top cup. The water vapor collected and cooled on the inside surface of the top cup causing it to *condense* to become liquid water. Since water vapor is invisible, the only way to compare the amount of water vapor is to compare the amount of condensation on the inside of the top cups. Tell students that evaporation and condensation were probably also happening in the cups with room-temperature water, but at a slower rate. Evaporation and condensation can occur at a wide range of temperatures, but it is the *rate* of evaporation and condensation that is most affected by temperature.

Teacher note: The amount of water vapor already in the air will also affect the rate of evaporation. At a given temperature, the more water vapor that is in the air, the slower the evaporation rate.

Activity 5.3
Cooling water vapor

Question to investigate

Does cooling water vapor increase the rate of condensation?

In the first part of this activity, students saw that hot water evaporates faster than room temperature water to produce more water vapor. They also saw that when water vapor cooled on the inside of the top cup, it condensed to form liquid water. In this activity, students will help design an experiment to answer the question: Does cooling water vapor increase the rate of condensation? Students will realize that they will need two samples of water vapor and a method of cooling one of them in order to investigate this question.

1. Discuss with students some of their own experiences with condensation.

Ask students questions like the following to get them thinking about condensation:

- Do you ever notice that if you have a cold drink on a warm humid day, the outside of the cup or can gets wet? Where do you think this moisture comes from?
- You may have made a cold window "cloudy" by breathing on it and then drawing on the window with your finger. Where did that cloudiness come from? Tell students that in both of these examples the water vapor from the air, or their breath, is cooled enough to change state to liquid water. This is called *condensation*.
- Another example of water vapor condensing is when the moist air in a cloud turns to rain. Have students consider the other condensation examples and ask them what might cause the water vapor in a cloud to condense to form rain. Students should say that the condensation is probably caused by cooling the water vapor.

2. Have a discussion about designing an experiment to find out: Does cooling water vapor increase the amount of condensation?

Ask students for their ideas of what they might need to do in an experiment to answer the question. Students should realize that they will need a sample of water vapor, some way to trap it, and some way to cool it. Then ask them for their ideas on ways they might be able to do this. Ask students how they might get or make a sample of water vapor. If students need more prompting, ask them how they made water vapor in the last activity.

Discuss with students how they might conduct the experiment. Let students know that while it is important to cool one sample of water vapor, they will need another identical sample of water vapor that they do not cool. Ask students why it is important to have this uncooled sample of water vapor. Students should realize that the uncooled sample will be the *control* and will allow them to see whether or not cooling water vapor increases condensation.

3. Have groups design and conduct their experiment.

The students' procedure may be like the following. It is fine if students have other ways to cool a sample of water vapor, like placing one trapped sample outside on a cold day or placing one in the refrigerator.

Procedure

1. Fill two punch cups about ⅔ full of hot tap water.
2. Quickly place a tall clear plastic cup over each of the punch cups as shown.
3. Place a piece of ice on the *top* of one of the cups and wait for 2–3 minutes.
4. After the ice has been on the cup for 2–3 minutes, remove it and use a paper towel to dry off the water from the melted ice.
5. Use your magnifier to look at the *top* of each cup.

4. Have students record their observations.

Pass out *Activity sheet 5.3—Cooling water vapor* on page 128 after students complete this experiment.

5. Have students share their experimental design and observations.

Have each group of students report their method of cooling water vapor and their observations.

Expected results: **The water droplets on the *top* of the cup with ice will be bigger than the droplets on the *top* of the cup without ice. The difference in droplet size may not be extreme but should be noticeable to students, especially when they use the magnifying glass. If the activity is done without the magnifying glass, the difference is not as obvious but should still be recognizable.**

Regardless of the method students used to cool water vapor, they should have noticed more condensation in the sample of cooled water vapor.

6. Ask students to draw conclusions, based on their experiments, about the effect of temperature on evaporation and condensation.

Ask students what they learned about *evaporation* and *condensation* in Activities 5.2 and 5.3. Also ask students what they used as a control in each experiment and why it was important to use them. Students should have learned that increasing the temperature increases the rate of evaporation and decreasing the temperature increases the rate of condensation. The purpose of the control in each activity was to provide a reference so that the effect of raising or lowering temperature could be compared.

Activity 5.4a
Explore student experiences with condensation: Moisture on the outside of a cold cup

Question to investigate

Where does the condensation that appears on the outside of a cold cup come from?

In this activity, students will explore what might cause the common observation of moisture forming on the outside of a cold cup. Students will discover that it is the water vapor in the air that condenses on the cup that causes the moisture. Since the amount of water vapor in the air varies during different times of the year, be sure to test this experiment before trying it with students.

Teacher preparation: Before trying this activity with students, test the air in your classroom for minimum adequate moisture levels by placing water and ice in a clear plastic cup and leaving undisturbed for 3–5 minutes. If moisture is readily observable on the outside of the cup, you can do this activity with your students. If not, try Activity 5.4b.

1. Discuss with students their experiences with moisture on the outside of a cold cup or other container.

Introduce this activity to students by telling them about a common example of condensation. In the summer, when they have a cold drink, the outside of the glass or can may get wet. Ask students if they have noticed this before and have them give a few examples. Tell students that they will do an experiment to see this phenomenon and to find out where the moisture comes from.

2. Compare the outside of a cup with ice cold water to the outside of a cup with room-temperature water.

You may choose to follow this procedure as a demonstration, or you may prepare these cups ahead of time and pass them out to each group so that students can observe them up close.

Procedure

1. Fill 1 cup with ice. Add water until the cup is about ¾ filled.

2. Add room-temperature water to another cup until it is about ¾ filled.

3. Wipe the outside of both cups with a paper towel to be sure they are dry.

4. Allow the cups to sit for about 3–5 minutes. Look at the outside of each cup. Use your finger to test for any liquid on the outside of the cups.

3. Discuss student observations.

Ask students what they observe on the outside of each cup.

Expected results: **Moisture will appear on the outside of the cup with ice water, but nothing noticeable will appear on the outside of the cup with room-temperature water.**

Then ask students the following questions to get them thinking about where the moisture comes from. It is not important that students know the correct answers to these questions at this point. They will explore these further in the next part of the activity.

- What do you think the liquid is on the outside of the cup?
- Where do you think the liquid comes from?
- Why do you think the cold cup has more moisture on the outside than the room-temperature cup?

4. As a class, design an experiment to find out where the moisture on the outside of a cold cup comes from.

Review students' ideas about where the liquid on the outside of the cold cup may have come from. Students might suggest that water could have leaked through the cup, it could have come from the air, or they may offer some other ideas. If students think that the water may have leaked through the cups, point out that water appeared only on the outside of the cold cups and not on the room-temperature cups. If the cups leaked, it is extremely unlikely that only the cups with cold water leaked and the cups with room-temperature water did not.

Suggest to students that their ideas about where the moisture came from may or may not be true. Tell them that they will help you design an experiment to investigate if the moisture they observed on the cold cups had something to do with the air surrounding the cups. Then, in groups, they will do the experiment.

Teacher note: Students may have some difficulty thinking of how to design a simple experiment to investigate the possible role of air in producing the moisture on the outside of the cold cup. The idea of exposing one cold cup to air and limiting the exposure of another cold cup to air, may be difficult for students to think of on their own. You may need to offer suggestions like those in the sample dialogue below. As you model the thinking processes involved in designing this experiment, students will better understand issues such as identifying and controlling variables and the use of a control.

Ask students questions like the following to lead the design of this investigation:

- Which cup had moisture on the outside, the one with cold water or the one with room-temperature water?

Since the cup with cold water had the moisture on it, that is what we will use to find out if air has anything to do with the moisture.

- How many cups of cold water should we use?

We will need more than one cup. Since we are trying to see if air has anything to do with the moisture on the cold cup, we will need one cold cup exposed to air and one cold cup *not* exposed to air. The only way to see if the air around a cold cup really makes a difference is to have one cup with air around it and one cup with no air around it so that we can compare them. The cold cup without air is your control. So we need two cups.

- How can we limit the amount of air that is allowed to touch the control cup?

Removing all the air from around the control cup would be very difficult to do in this classroom, so we will take away as much air as we can by putting the control cup in a plastic bag and squeezing out as much air as possible.

- To make this a fair test, what will we have to keep the same in both cups?

The cups themselves should be the same, and both should have the same amount of ice and water.

5. Have groups conduct the experiment.

Procedure

1. Fill 2 cups with ice. Then add water until each cup is about ¾ full. Wipe the outside of both cups with a paper towel to be sure they are dry.

2. Carefully place 1 cup in a zip-closing plastic bag. Get as much air out of the bag as possible and then seal the bag tightly.

3. After about 3–5 minutes, observe the 2 cups. Does one have more moisture on the outside than the other?

6. Have students discuss their results.

Ask students which cup had more moisture on its outside.

Expected results: **The cup outside of the bag will have condensation on it while the cup in the bag will not. Since a much larger amount of air was around the cup outside the bag, students should conclude that the moisture on the outside of the cup had something to do with the air.**

Have students use their observations from this experiment and others within this investigation to help explain where the moisture may have come from. Students should conclude that the moisture came from water vapor in the air that condensed on the outside of the cup.

7. Have students apply their learning to three different hypothetical situations.

Pass out *Activity sheet 5.4—Explore student experiences with condensation* on page 129.

Activity 5.4b
Explore student experiences with condensation: Moisture on the outside of a cold cup—For dry environments

Question to investigate

Does the amount of water vapor in the air have an effect on whether moisture forms on the outside of a cold cup?

If the air is not moist enough to get observable condensation in activity 5.4a, this activity can be used because students create the moist air as part of the activity. Students use cold film canisters to discover that a certain amount of water vapor in the air is required to produce moisture on the outside of a cold cup.

1. Do a demonstration showing that moisture does not form on the outside of a cold cup in a dry environment.

Teacher preparation:

- Fill a cup with ice. Add water until the cup is about ¾ filled.
- Allow the cup to sit for at least 3 minutes.

Introduce this activity to students by telling them about a common example of condensation. In the summer, when they have a cold drink, the outside of the glass or can may get wet. Ask students if they have noticed this before and have them give a few examples.

Show students the cold cup that you have prepared and point out that there is no noticeable moisture on the outside of the cup. Ask students what might be the reason that this cup does not have moisture on the outside, when they have seen it on the outside of cold cups at other times. Students may suggest that it is not warm enough in the classroom, that the ice water is not cold enough, that moisture might not appear on plastic but will on a glass or can, that moisture only appears on a cup outdoors, that there needs to be more humidity in the air, or other ideas. These suggestions are good and could lead to fruitful investigations, either in class, at home, or as science fair projects. Tell students that this activity will investigate whether the classroom air does not have enough water vapor in it to form moisture on the outside of a cold cup.

2. Have students help design the experiment.

One sample will be classroom air.
Now that students know they will investigate whether the low amount of water vapor in the classroom air is preventing moisture from forming on the outside of a cold cup, ask students for their ideas about how to set up an experiment to test this hypothesis. This may take some prompting, but students should suggest that they will need a sample of classroom air and a sample of air that has a higher amount of water vapor.

Decide how to make one sample with more water vapor in it.
Ask students how they could increase the amount of water vapor in the air. Students may suggest bringing in a humidifier, running hot, steamy water from the sink, letting a tub of hot water evaporate, or other ideas. These ideas might work, and could be tried. The procedure below describes one way to create the sample of air with a higher amount of water vapor and to investigate the question.

Decide to use identical cold containers in the sample of air with low water vapor and in the sample of air with higher water vapor.
Once students realize they need a sample of air with low water vapor and a sample of air with higher water vapor, ask them what they should place in each sample. Since this experiment is meant to see the effect of different amounts of water vapor on a cold cup, students should suggest placing a cold cup in each sample of air. If students make a sample of air with high water vapor, the way they did in Activity 3, they will need to use a small container like a film canister for the cold cup as suggested in the procedure below.

3. Conduct the experiment.

Procedure

1. Fill a clear plastic punch cup about ⅔ full with hot tap water.

2. Immediately place a taller plastic cup upside down on top of the punch cup.

3. Fill 2 plastic film canisters with ice. Add water until they are nearly full and snap the caps in place.

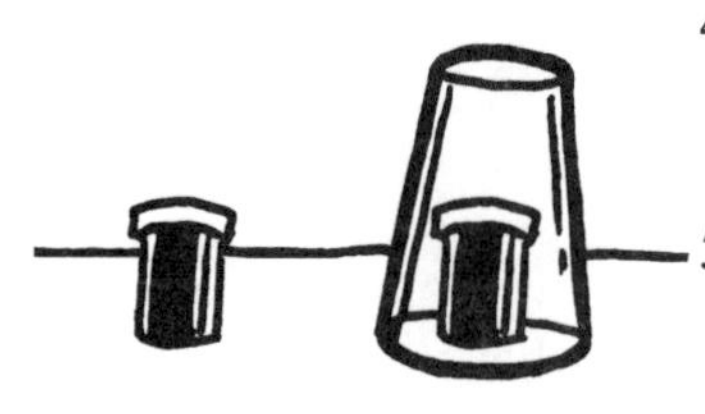

4. Wipe the outside of both film canisters with a paper towel to be sure they are dry.

5. Once the tall cup appears cloudy, take it off of the punch cup and immediately place it over one of the film canisters. Wait 2–3 minutes.

6 Remove the tall cup and look at the outside of each canister closely. Touch the outside of each canister with your finger.

4. Discuss student observations.

Ask students if they notice any difference on the outside of each film canister.
Expected results: **Moisture appears on the outside of the film canister that is placed under the cup with a lot of water vapor in it. The film canister that was exposed to the classroom air does not have noticeable moisture on it.** Ask students if the hypothesis was right: that the classroom air didn't have enough water vapor in it to cause moisture to form on the outside of a cold cup. Students should agree that their observations support this hypothesis. Ask students what might cause the moisture to form on the outside of a cold cup in the summertime. Students should realize that when they have seen moisture appear on the outside of a cold cup, there must have been more water vapor in the air than there is in the classroom.

5. Have students apply their learning to three different hypothetical situations.

Pass out *Activity sheet 5.4—Explore student experiences with condensation* on page 129.

Activity 5.4c
Explore student experiences with condensation: Breathing on cold windows to make them "cloudy"

Question to investigate

What causes some surfaces to look "cloudy" when you breathe on them?

In the wintertime, students may have used their warm moist breath to "cloud up" a car or house window in order to write or draw with their finger. Students may also have noticed that they can sometimes see "smoke" when they talk or exhale in winter. In this activity, students will breathe on the outside of a cold cup and see the "cloudiness." Based on their experience with condensation from previous activities, students can conclude that these wintertime effects are a result of condensation.

1. Discuss with students their experiences with using their breath to make a cold window "cloudy."

Introduce this activity by asking students if they have ever breathed on a cold window to make it "cloudy." Tell students that they will do an experiment to try to find out what causes this cloudiness.

2. Conduct an experiment to see if the temperature of the surface has something to do with the "cloudiness."

As you explain to students that they will breathe on the outside of a cold cup and a room-temperature cup, ask them to predict what they might see on the outside of each cup. Students should have enough experience with condensation to predict that the outside of the cold cup will have more moisture on it and will look more cloudy.

Procedure

1. Fill a cup with ice. Add water until the cup is about ¾ full. Place ¾ cup of room-temperature water in another cup.

2. Wipe the outside of both cups with a paper towel to be sure they are dry.

3. Slowly breathe warm air from your mouth onto the outside of the room-temperature cup and then the cold cup.

4. Use your finger to feel the outside of each cup.

3. Have students report their observations.

Expected results: **As student breath hits the side of the room-temperature cup, it becomes slightly cloudy and then quickly clears. As student breath hits the side of the cold cup, the side becomes cloudy and stays that way longer. When students feel each of the cups, they should notice that there is moisture on the outside of the cold cup.**

Ask students what they think the process was that caused the moisture on the outside of the cup. Based on Activity 5.3, students should conclude that the process is condensation. Ask them what two things are needed to cause or increase condensation. Students should say that water vapor is needed to cause condensation and that the colder the temperature, the more condensation there is. Ask students where the water vapor came from in this experiment to cause the condensation. Students probably will realize that the water vapor comes from their breath. Ask students to relate this experiment to their experience of breathing on a cold window to make it "cloudy." They should realize that water vapor from their breath condenses on the cold window. Ask students what might cause their breath to look like "smoke" when they breathe out on a cold winter day. Again they should realize that this is a result of condensation. The water vapor in their breath turns to tiny droplets of liquid water in the air. The way the light hits these droplets makes them look like "smoke," a "cloud," or "steam."

4. Have students apply their learning to three different hypothetical situations.

Pass out *Activity sheet 5.4—Explore student experiences with condensation* on page 129.

Activity 5.5
From gas to liquid to solid

Question to investigate

How can you make frost appear on the outside of a can?

This activity is an extension of Activity 5.4a in which ice was used to make the container cold. Here, a metal can is used and salt is added to make the container even colder. Water vapor from the air condenses on the can to form liquid water and then *freezes* to become solid frost. The method of adding salt to ice to cool the water vapor in the air is similar to the method used in homemade ice cream makers. As in Activity 5.4a, this activity will only work if there is sufficient water vapor in the air. Students have already learned that water vapor from the air can condense to become liquid water. In this activity, they will see that water can change state again and freeze to become ice.

Teacher preparation: Obtain one standard-sized soup can for each group and remove the label. If you see any sharp or jagged edges around the top, carefully press them down with pliers and cover the rim with duct tape.

1. Discuss how cooling can cause changes in state.

Tell students that they found out that cooling water vapor enough causes it to change state from a gas to a liquid. Ask students how they could get water to change state again from a liquid to a solid. Students should suggest making the liquid even colder until it turns to ice. Tell students that they will do an activity to see if they can cool water vapor enough to get it to change to a liquid and then to a solid.

2. Add ice and salt to a can to create a very low temperature.

Procedure

1. Dry the outside of a can with a paper towel.
2. Place 3 heaping teaspoons of salt in the bottom of the can. Fill the can about halfway with crushed ice.
3. Add another 3 heaping teaspoons of salt.
4. Add more ice until the can is almost filled and add another 3 teaspoons of salt.
5. Hold the can near the top and mix the ice/salt mixture with a pencil for about 1 minute. Remove the pencil and observe the outside of the can. Do not touch it.

Teacher note: After completing step 5, you may choose to have students place a thermometer inside the can because the temperature of the salt and ice mixture will be below the normal freezing point of water, which is 0 °C. The reason for this subfreezing temperature is explained in the teacher background information.

3. Have students record their observations.

Pass out *Activity sheet 5.5—From gas to liquid to solid* on page 130 as students complete this activity. They will need to record their observations with a drawing so it is best if the can is still available to them.

4. Discuss student observations.

Ask students what they notice on the outside of the can. Have them touch the outside of the can and report what they observe.

Expected results: **The outside of the can will become covered with a thin layer of frost. Students will notice this frost on the coldest part of the can at or below the level of the ice. Above the ice, the can is cold, but not cold enough to change the moisture on the outside of the can to frost.** Ask students why they think there are water droplets on the top of the can and frost on the bottom of the can.

Now that students have been through the series of activities on states of matter, ask them how water vapor changes when the temperature is lowered. Students may remember from Activity 5.1 that cooling a gas causes it to contract. Since water vapor is a gas, cooling causes water vapor to contract. Students should say that lowering the temperature enough will make water vapor *condense* to liquid water and that lowering the temperature of the water enough again will cause it to *freeze* to form ice.

Students have described changes in state based on cooling water vapor to a liquid and then cooling the liquid to a solid. Now have students think about and describe the opposite process. Ask students how heating the frost on the outside of the can could cause it to change state all the way back to water vapor. Students should say that increasing the temperature enough will cause the ice to *melt* to liquid water and that increasing the temperature of the water enough again will cause it to *evaporate* to form water vapor.

Name: ____________________

Activity sheet 5.1

Temperature affects a gas

What was inside the "empty" bottle you used during this experiment? ____________________

What state of matter is this substance? ____________________ (Hint: solid, liquid, or gas)

Draw before-and-after pictures of the bottle to show what the bottle looked like *before* you warmed it in the hot water and what it looked like *after* you warmed it.

before	*after*

What caused the bubble to look like it does in your *after* picture?

____

____

What happens to a gas when it is heated?

____

____

____

What happens to a gas when it is cooled?

____

____

Name:________________________________

Activity sheet 5.2

Evaporation and condensation

	Matter changes state *from* a...	*to* a...
Melting	solid	liquid
Condensation		
Evaporation		

Which produced more water vapor, the hot water or the room-temperature water?

____

Water vapor is invisible, so how could you tell which temperature of water produced more water vapor?

____

____

Name:________________________________

Activity sheet 5.3

Cooling water vapor

Draw and label what you observed on the top of each tall cup.

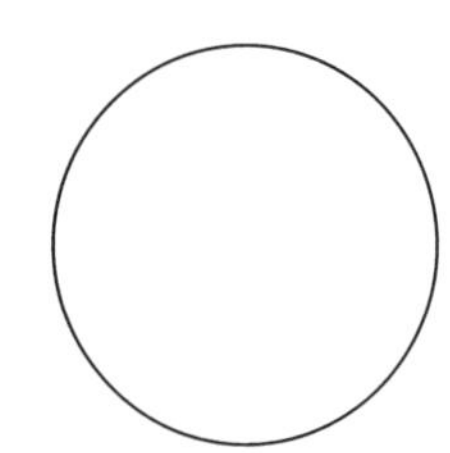

Cup that had ice on it

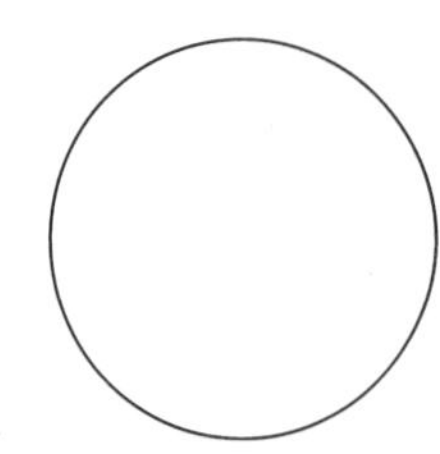

Cup that didn't have ice on it

How did the hot water and cold ice cube affect evaporation and condensation?

____

____

____

____

Name:______________________________

Activity sheet 5.4

Explore student experiences with condensation

1. Why wouldn't you observe moisture on the outside of a cup filled with cold water in a very dry environment like the desert during the dry season?

____

____

____

____

____

2a. In very cold weather, you may notice that your breath looks like "smoke" when you talk or breathe out. Is this "smoke" a result of condensation? ______________

2b. What is this "smoke" made of?

____

____

____

____

____

3. If you were in an environment that had very warm humid air, like a tropical rain forest, and you had a cold glass of lemonade, you would see tiny droplets on the outside of the glass. Why do you think you wouldn't see as many, if any, tiny water droplets on the outside of the glass if you were drinking hot chocolate in this same environment?

____

____

____

____

____

Name:______________________________

Activity sheet 5.5

From gas to liquid to solid

Label the states of matter that you observed on the outside of the can as you stirred the ice and salt.

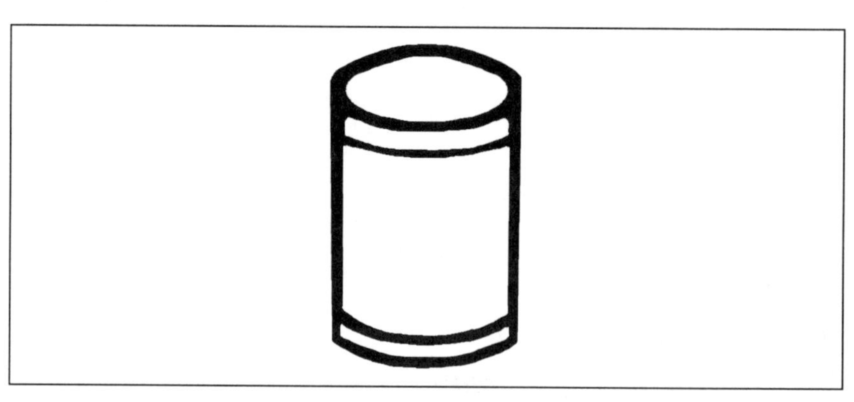

Explain the relationship between changes in temperature and changes in state.

____

____

____

	...heated?	...cooled?
What change in state might happen to a **gas** that is...		
What change in state might happen if a **liquid** is...		
What change in state might happen to a **solid** that is...		

Name:____________________________

Assessment rubric

Investigation 5—States of matter

To earn a "B," a student must receive a "Very good" in each category.

	Very good	Satisfactory	Needs improvement
Activity sheet 5.1	______	______	______

Temperature affects a gas

Records observations drawings
Explains effect of heating and cooling a gas

	Very good	Satisfactory	Needs improvement
Activity sheet 5.2	______	______	______

Evaporation and condensation

Identifies state changes in melting, evaporation, and condensation
Uses evidence from observations to explain the effect of temperature on water vapor

	Very good	Satisfactory	Needs improvement
Activity sheet 5.3	______	______	______

Cooling water vapor

Records observations with detailed drawings
Uses evidence from observations to explain the effect of temperature on evaporation and condensation

	Very good	Satisfactory	Needs improvement
Activity sheet 5.4	______	______	______

Explore student experiences with condensation

Applies learning from the activities to hypothetical situations

	Very good	Satisfactory	Needs improvement
Activity sheet 5.5	______	______	______

From gas to liquid to solid

Records observations with a labeled drawing
Describes relationship between temperature and state changes

	Very good	Satisfactory	Needs improvement
Investigative behaviors	______	______	______

Participates in design of tests
Conducts fair tests
Participates in class discussions
Works cooperatively with group
Uses evidence to formulate explanations

To earn an "A," a student must also exhibit some of the following qualities throughout this investigation.

Shares detailed observations
Participates well in class discussions
Participates well in group work
Uses scientific thinking
Consistently exhibits exceptional thought and effort in tasks

Investigation 6. Density

Summary

The density of an object and whether it *sinks* or *floats* are observable properties of an object that are familiar to children. Trying to understand the nature of sinking and floating and their connection to density can lead to many opportunities for observation, measurement, and designing and conducting investigations. Students will weigh equal volumes of different substances and compare them with the weight of an equal volume of water to predict whether the substance will sink or float.

Objective

Students will conduct investigations to discover which objects or liquids float or sink in water. They will compare the relative densities of these objects and liquids by weighing equal volumes of each with a balance and comparing the result to the weight to an equal volume of water. Students will use their results to predict whether a given substance will float or sink.

Assessment

The assessment rubric *Density* on page 157 is included so that you can assess and document student progress throughout the investigation. The abilities and understandings demonstrated by students and recorded on the rubric include the following: As students weigh equal amounts of wax, clay and water, they will begin to develop a meaning of density. They should discover that it is necessary to consider both the density of an object and the density of a liquid when predicting whether or not an object will sink or float in that liquid. Students should realize that density is a function of both mass and volume and that dissolving a substance in a liquid can change the density of the liquid and that changing the volume of an object can change the density of the object. Investigative behaviors observed as students use a balance scale, help design experiments, and use evidence from their experiments to formulate explanations are also recorded on the scoring rubric.

The end of *Activity 6.2—Comparing the density of different liquids* and *Demonstration 6b—Applying density to sink and float* can both be used as assessments. On Activity sheet 6.2, students will infer the relative densities of four different objects and three different liquids based on the way they layer in a cup. On Demonstration Sheet 6b, students will use what they have learned about density in the investigation to explain why a heavier block of wood floats in water while a lighter pebble sinks.

Relevant *National Science Education Standards*

Physical Science

K–4

Properties of Objects and Materials

Objects have many observable properties, including size, weight, and shape.
Those properties can be measured using tools, such as rulers and balances.
Objects can be described by the properties of the materials from which they are made, and those properties can be used to separate or sort a group of objects or materials.

5–8

Properties and Changes of Properties in Matter

A substance has characteristic properties such as density…all of which are independent of the amount of the sample.

Science as Inquiry

K–4

Abilities Necessary to do Scientific Inquiry

Ask a question about objects.
Plan and conduct a simple investigation.
Employ simple equipment and tools to gather and extend the senses.
Use data to construct a reasonable explanation.
Communicate investigations and explanations.

Understandings about Scientific Inquiry

Scientific investigations involve asking and answering a question.
Scientists use different kinds of investigations depending on the questions they are trying to answer.
Types of investigations include describing objects…and doing a fair test.
Good explanations are based on evidence from investigations.

5–8

Abilities Necessary to do Scientific Inquiry

Identify questions that can be answered through scientific investigations.
Design and conduct a scientific investigation.
Use appropriate tools and techniques to gather, analyze, and interpret data.
Develop descriptions, explanations, predictions, and models using evidence.
Think critically and logically to make the relationships between evidence and explanations.
Communicate scientific procedures and explanations.

Understandings about Scientific Inquiry

Different kinds of questions suggest different kinds of scientific investigations.
Scientific explanations emphasize evidence and have logically consistent arguments.
Scientific investigations sometimes result in new ideas and phenomena for study or generate new procedures for an investigation. These can lead to new investigations.

How this investigation relates to the *Standards*

One goal of the *Standards* is for students to understand that substances have characteristic properties and that these properties remain the same regardless of the size of the sample. Through the following investigation, students will discover that density is a characteristic property of a substance and determines whether or not a substance will float or sink in another substance.

Materials chart

6.1 Comparing the density of an object to the density of water
6.2 Comparing the density of different liquids
6.3 Changing the density of a liquid
6.4 Changing the density of an object

Each group will need

Each group will need	Activities 6.1	6. 2	6.3	6.4
water in clear cup	•	•	•	•
pencil	•	•		
masking tape	•	•		
marker	•	•		
metric ruler, 30 cm	•	•		
tea light candles with their metal containers		2		
clay balls, about 3½ cm in diameter	1			2
tall clear plastic cups		1	1	
paperclips		~50		
vegetable oil in cup		•		
corn syrup (Karo) in cup		•		
3½-ounce cups		4		
plastic teaspoon			•	
salt			•	
carrot slice			•	
crayon piece		•		
toothpick or Popsicle piece		•		
pasta piece		•		

Notes about the materials

The paperclips will be used with a balance to weigh different substances. Use standard-sized metal paperclips.

Teacher preparation

Teacher will need for the demonstrations

Demonstration 6a—Introducing density through sink and float
6.3 Changing the density of a liquid
6.4 Changing the density of an object
Demonstration 6b—Applying density to sink and float

	Demo 6a	Activities 6. 3	Activities 6.4	Demo 6b
small pebble	•			•
block of wood	•			•
balance	•			•
large clear container	•			•
water	•	•	•	•
clear plastic cups		3	1	
salt		•		
carrot slices		3		
plastic teaspoon		•		
clay ball			•	

Notes about the materials

The block of wood must be heavier than the pebble.

Activity sheets

Activity sheet 6.1 page 153

Comparing the density of an object to the density of water

Copy one per student.

Students will use this activity sheet to record their observations as they do the activity. They also will use their experience with density from this activity to explain why an ice cube floats in water.

Activity sheet 6.2 — page 154
Comparing the density of different liquids

Copy one per student.

Students will compare the weights of equal amounts of water, vegetable oil, and corn syrup. They will use these weights to infer the relative densities of the liquids. Students will also discuss the relative densities of the liquids and objects in a cup as they explain why they arrange themselves at various levels.

Activity sheet 6.3 — page 155
Changing the density of a liquid

Copy one per student.

On this activity sheet, students will record the method their group used to make a carrot slice "hover" in a cup of salty water. They will also explain the relationship between density and sinking and floating in the context of this activity.

Demonstration sheet 6b — page 156
Applying density to sink and float

Copy one per student.

Along with this activity sheet, the teacher will once again show students the demonstration from the beginning of the investigation. Students will then use what they have learned about density to explain why the heavy wood floats and the lighter pebble sinks.

Science background information

The *density* of a substance is defined as its mass divided by its volume. One way to compare the densities of two substances is to weigh equal volumes of each substance. If substance A weighs more than an equal volume of substance B, then substance A is more dense than substance B, and substance B is less dense than substance A. Since density is the result of the mass of the substance and how much space it occupies, it is a fundamental or characteristic property of a substance. There is a common misconception that the density of a substance is determined only by how closely packed or squeezed together the particles are that make up the substance. This is only part of the story. The mass of the particles is very important too. It is possible that a substance with very heavy particles spaced a little farther apart could have a higher density than a substance whose particles are lighter but packed more closely together. So the density of a substance is a result of the mass of its particles and how they are arranged in the substance.

The topic of sinking and floating is directly related to density. Whether an object floats or sinks in a fluid depends on the density of the object and the density of the fluid. If the object is more dense than the fluid, the object will sink. If it is less dense than the fluid, the object will float. The density of a substance is not affected by the size of the sample. For instance, wood is a substance that is less dense than water. A small piece of wood floats and so does a very large piece, like a tree trunk. If a substance is more dense than water, it sinks, regardless of the size of the sample. For example, rock is more dense than water. A huge boulder will sink and so will a tiny grain of sand.

Demonstration 6a—Introducing density through sink and float

Students see that the wood weighs more than the rock, but the wood floats and the rock sinks. These observations may seem counterintuitive to students. After a discussion of the factors that might be causing the heavy wood to float and the lighter rock to sink, students should be guided to begin to consider that "density" has something to do with this phenomenon. They should also be introduced to the idea that both the density of the objects and the density of the water matter in sinking and floating.

Activity 6.1—Comparing the density of an object to the density of water

This activity shows that to determine whether an object will sink or float, the weight of the object is compared to the weight of an equal volume of water. Students find that wax weighs less than an equal volume of water. This means that wax is less dense than water and should float. Since clay weighs more than an equal volume of water, it is more dense than water and should sink.

Activity 6.2—Comparing the density of different liquids

Students will need to compare the weights of equal volumes of the different liquids. From this information, they can determine the relative densities of all the liquids and order them from most dense to least dense to make a density tower.

Activity 6.3—Changing the density of a liquid

Since the carrot sinks in fresh water, it must be more dense than fresh water. Dissolving salt in water increases the mass of the solution but doesn't increase its volume much at all. Since density equals mass divided by volume, and adding salt to water increases the mass of the salt water solution, the density of salt water is greater than the density of fresh water. When enough salt is dissolved in the water, the liquid becomes more dense than the carrot, and the carrot floats.

Activity 6.4—Changing the density of an object

The density of the clay itself is not changed but the density of the overall object is altered by changing the shape of the object. Since density = mass/volume, increasing the volume of an object without increasing its mass must decrease its density. The volume of these clay boxes = length x width x height. One or more of these dimensions can be increased to increase the volume and decrease the density. The trick is to increase the volume enough so that the corresponding decrease in density will make the clay box less dense than water so it will float.

Demonstration 6a
Introducing density through sink and float

Question to investigate

Do heavy things sink and light things float?

When a pebble and a piece of wood are placed on a balance, the wood is shown to be heavier than the pebble. However, when both are placed in water, students will see that the wood floats and the pebble sinks. This observation provides a starting point for a discussion about density and its relationship to sinking and floating.

1. Discuss with students their experience with objects that sink and objects that float.

As a whole class, in small groups, or as a simple homework assignment, have students brainstorm a list of objects that they know sink and a list of objects that they know float. During a class discussion, ask students if they notice anything that the objects in each list have in common. Students may find that many objects that sink are heavy and many objects that float are light. However, there will likely be exceptions, like aircraft carriers and large logs. Keep this list to use as an assessment at the end of this investigation.

2. Demonstrate that although the wood is heavier than the pebble, the wood floats and the pebble sinks.

Procedure

1. Place the pebble on one end of a balance scale.

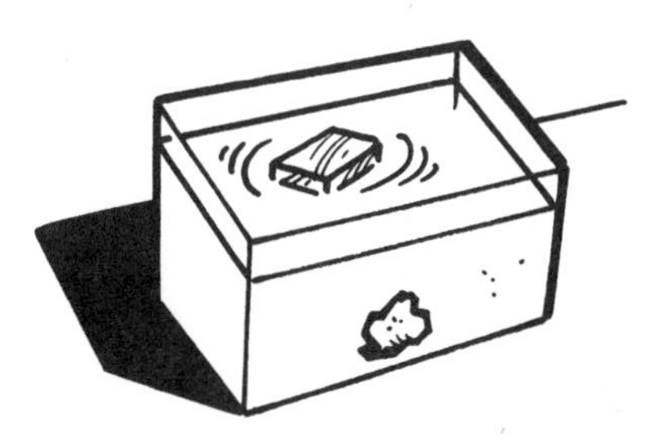

2. Place the block of wood on the other end so that it is obvious that the wood is much heavier than the pebble.
3. Then place the pebble in the water and let it sink.
4. Place the wood in the water and let it float.

3. Discuss student observations and introduce the term density.

Expected results: **The wood floats and the pebble sinks.** Ask students how something light like the pebble sinks while something heavier like the piece of wood floats. Students might state facts such as “wood floats” and “rocks sink.” Some students might say that rock is heavy for its size and wood is light for its size. This answer is moving in the right direction because it is getting closer to an explanation based on the density of the pebble and the wood. Try to get students to understand that even though the pebble is light, its weight is packed into a small amount of space. Even though the wood is heavier, its weight is spread out over a larger amount of space.

Tell students that this relationship between the weight of an object and the amount of space that it takes up is called the *density* of the object. Explain that if they compare the weight of two objects that are the *same size and shape* (take up the same amount of space), the one that is heavier is *more dense* and the one that is lighter is *less dense*. Suggest to students that they imagine a piece of rock and a piece of wood that are the same size and shape and decide which would be heavier. Then ask students which is more dense, the rock or the wood.

At this point, students have been introduced to the concept of density. They should also be able to say that the rock is more dense than the wood. However, students still have not dealt with why the rock sinks and the wood floats. This question has to do with the relationship between the density of the object and the density of water, which will be explored in the next activity. Ask students if based on their observation of the rock and wood they can say that heavy things sink and light things float. They should be able to conclude that heavy things don't always sink and light things don't always float. Students should realize that the density of the object seems to have something to do with sinking and floating. Let them know that in the next activity, they will explore this idea.

Activity 6.1
Comparing the density of an object to the density of water

Question to investigate

How can you predict whether an object will sink or float in water?

Whether an object will sink or float in water depends on the density of the object compared to the density of water. One way to compare the density of an object to the density of water is to weigh equal volumes of each. In this activity, students will compare the weight of a wax candle to the weight of an equal volume of water. Students will discover that since the wax weighs less than an equal volume of water, it is less dense than water and will float. They will then compare equal volumes of clay and water to discover that since the clay weighs more than an equal volume of water, it is more dense and will sink.

1. Have students construct a simple balance scale.

Tell students that in order to predict if an object is going to sink or float in water, they will have to know about the density of the object compared to the density of water. Tell students that a balance scale, like the kind they will make, will help them compare these densities.

Procedure

1. Tape the pencil down as shown. Roll two small pieces of tape so that the sticky side is out. Stick each piece of tape to the opposite ends of the ruler.

2. Remove both candles from their metal containers. Place an empty metal container on each piece of tape. Be sure that the edge of the metal container lines up with the end of the ruler as shown.

3. Lay the ruler on the pencil so that it is as balanced as possible. The spot on the ruler directly above the center of the pencil is your *balance point.* Mark the ruler with a pencil or permanent marker at this point.

2. Have students compare the weight of equal volumes of wax and water.

Pass out *Activity sheet 6.1—Comparing the density of an object to the density of water* on page 153, so that students can record their observations as they do the activity.

Procedure

1. Carefully place one of the candles back into its metal container on one end of the ruler. Make sure the same balance point is directly over the center of the pencil.
2. Then, carefully pour water into the metal container on the other end of the ruler. Be sure to fill the container with water to the same height as the wax fills the other container.

Teacher note: Some tea light candles are a little thicker in the middle than around the edges. When using the edge of the candle to judge how much water to add to the other candle container, this extra volume of wax should be taken into account and a little extra water should be added to attempt to match the actual volume of the wax.

3. Discuss student observations.

Ask students the following questions:

- Which weighs more, the wax or an equal volume of water?
- Which is more dense, wax or water?

Expected results: **The water weighs more than the equal volume of wax and is therefore more dense than wax.**

4. Ask students to predict whether wax will sink or float in water.

Once students have made a prediction, ask them to explain it. Then have them test their prediction by placing the wax in a cup of water.

Expected results: **Since the wax is less dense than water, students should be able to predict that the less dense wax will float in the more dense water.**

Explain to students that the *density* of an object has two parts: the volume of the object and how much the object weighs. Tell them that they weighed the same volume of wax and water. Since the wax weighed less than an equal volume of water, the wax is less dense. When students put the wax in water, they noticed that it floated. Objects that are less dense than water will float. The density of an object compared to the density of water will determine whether or not an object will sink or float.

5. Have students determine whether clay will sink or float in water.

As a class, have students suggest methods to compare the weight of clay to the weight of an equal volume of water. They can suggest the same method they used to compare wax to water or another method that compares the weight of equal volumes of clay and water. Then, in their groups, have students follow one of these methods. Following is a sample procedure.

Procedure

1. Set up the ruler balance with the empty metal containers on each end. Check your balance point.
2. Fill one metal container with clay and replace it on the end of the ruler. Make sure the balance point is centered on the ruler.
3. Slowly and carefully add water to the empty container until it is filled.

6. Discuss students' observations and ask them to predict whether clay will sink or float in water.

Ask students questions like the following:

- Which weighs more, the clay or an equal volume of water?
- Do you think clay will sink or float in water? Why?

Have students test their prediction by placing the clay in a cup of water.

Expected results: **The clay weighs more than an equal volume of water. Since the clay is more dense than the water, students should be able to predict that the more dense clay will sink in the less dense water.**

Teacher note: If some students suggest that the clay could be made to float by changing its shape, tell them that they will investigate the effect of changing an object's shape in another activity.

7. Have students explain, in terms of density, why the objects they listed before the demonstration sink or float.

Examples:

Tree trunk
A tree trunk will float because wood is less dense than water. If you could weigh a large amount of water that has the same volume as the tree trunk, the tree trunk will weigh less than the water and will float.

Pebble
A pebble will sink because rock is more dense than water. If you could weigh a small amount of water that takes up the same amount of space as the pebble, the pebble will weigh more than the water and will sink.

Students should realize that if an object weighs more than an equal volume of water, it is more dense and will sink; and if it weighs less than an equal volume of water, it is less dense and will float.

Activity 6.2
Comparing the density of different liquids

Question to investigate

How do the densities of vegetable oil, water, and corn syrup help them to form layers in a cup?

Students will pour vegetable oil, water, and corn syrup in any order into a cup and discover that regardless of the order they are poured, the liquids layer the same way. Students will then weigh the liquids and use their results along with what they understand about density to explain why the liquids layer as they do.

1. Have students pour the three liquids in a cup in any order they choose.

Tell students to slowly and carefully pour the water, corn syrup, and vegetable oil into a clear plastic cup. Then have groups compare their results.

Expected results: **Corn syrup will sink to the bottom, water will be in the middle, and vegetable oil will float on the top.** Students will notice that regardless of the order in which the liquids are poured, they will arrange themselves in the same way.

2. Ask students what their results tell them about the density of each of the liquids.

Point out that the water is in the middle and that the oil *floats* on the water and that the corn syrup *sinks* in the water. Ask students, based on their experience with sinking and floating, what this means about the density of oil compared to the density of water, and about the density of corn syrup compared to the density of water. Students should realize that the vegetable oil is less dense than water and that the corn syrup is more dense than water.

3. Discuss with students how they could compare the weight of *equal* volumes of the liquids.

Ask students if they were to weigh equal volumes of the three liquids, which they would expect to be the heaviest, lightest, and in between. Since students know the relative densities of the liquids based on the way they layer in the cup, they should realize that if they weigh equal volumes of the liquids, corn syrup should be the heaviest, vegetable oil the lightest, and water in between. Ask students how they might go about weighing equal volumes of the liquids. Students may make or use a balance scale, like the one constructed in Activity 6.1, to compare equal volumes of the liquids on each side of the scale.

You could also suggest another method to students, in which they weigh each liquid against nonstandard units like paperclips, or some other unit. This method is described in the procedure below.

4. Have students compare the weights of equal amounts of the liquids.

Give students *Activity sheet 6.2—Comparing the density of different liquids* so that they can record which liquid is heaviest, lightest, and in between as they discover it. However, students should not finish this activity sheet until they have completed all of Activity 6.2.

The following procedure has students measure equal amounts of each liquid by marking 1 cm up on a small cup and pouring the liquids directly into the marked cups. Using a spoon to measure equal amounts of these liquids is not accurate because vegetable oil and corn syrup tend to stick to the spoon.

Procedure

1. Use a permanent marker to label 3 small cups *vegetable oil*, *corn syrup*, and *water*. Use your ruler to measure 1 cm up from the bottom of the cup and make a line with the marker.

2. Tape the pencil down as shown.
   Roll 2 small pieces of tape so that the sticky side is out.
   Stick each piece of tape to the opposite ends of the ruler.

3. Place the vegetable oil cup on one piece of tape and the empty unlabeled cup on the other. Lay the ruler on the pencil so that it is as balanced as possible. Use a pencil or permanent marker to mark the spot on the ruler directly above the center of the pencil. This is the *balance point*.

4. Very carefully add vegetable oil to its labeled cup until the oil reaches the mark on the cup.

5. Add paper clips, one at a time to the empty cup on the other end. Count the paper clips until the weight of the paper clips causes the oil cup to just lift from the table.

6. Repeat this process to see how many paper clips the same volume of water and the corn syrup each weigh.

5. Have students discuss their results.

Depending on the paperclips students used and the amount of liquid poured in each cup, students' results may vary a bit. However, it should be clear that the vegetable oil weighs less than the water and that corn syrup weighs more than the water.

Ask students questions like the following:
- Why is it important to weigh *equal volumes* of each liquid?
- Do your results from weighing the liquids agree with your observation of the layered liquids?

Expected results:

Liquid	Weight in paper clips
vegetable oil	24
water	29
corn syrup	41

6. Have students place a crayon piece, paper clip, piece of pasta, and toothpick or Popsicle piece into the cup of liquids.

When students place these objects in the layered liquids, the objects will position themselves as shown. Ask students to explain, in terms of density, why the objects end up where they do.

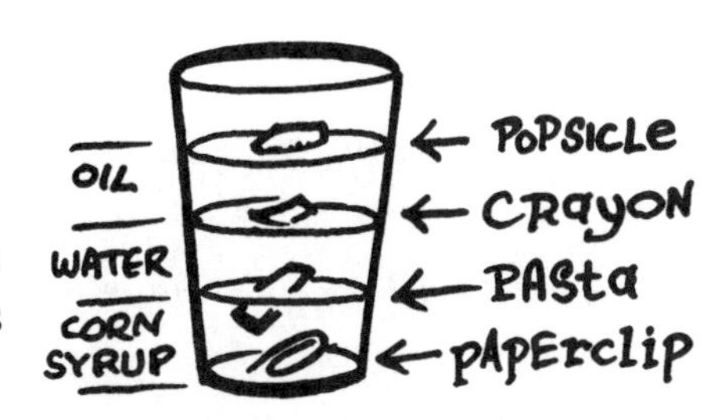

Activity 6.3
Changing the density of a liquid

Question to investigate

Can the density of a liquid be changed?

In this activity, students will add salt to water to show that the density of a liquid can be changed. Students will also discover that the density of a liquid can be changed so that an object will sink or float in the liquid.

1. Do a demonstration to show students one carrot slice sinking in a cup of fresh water and another floating in a cup of salt water.

Procedure

1. As students watch, pour ⅔ cup of fresh water into 2 clear plastic cups. Place 1 similar-sized carrot slice in each cup.

2. Add about a tablespoon of salt to one of the cups and stir. The carrot should begin to rise. Continue adding salt and stirring until the carrot floats to the top and stays there.

2. Have students discuss their observations and explain them in terms of density.

Expected results: **The carrot slice in the salt water will float, while the carrot slice in the fresh water will sink.** Ask students what might be causing the carrot to float in the salt water. Students should suggest that the reason the carrot floats has something to do with the salt. Ask students what they would expect if you placed equal volumes of water and this salt water on either end of a balance scale.

3. Do a quick demonstration to confirm the idea that the saltwater is more dense.

Depending on the type of balance scale you have, either place the cups of water and salt water directly on either end of the scale, or pour equal volumes of each into identical containers with lids and seal before placing them on either end of the scale.

Expected results: **The sample of salt water will weigh more than the equal volume of fresh water and is therefore more dense.**

4. Ask students what this demonstration tells them about density.

Ask students questions like the following:

- Before I added salt to the water, how did the density of the carrot compare with the density of the water?

Since the carrot sunk, the density of the carrot must be greater than the density of the fresh water.

- What happened to the density of the water as salt was added?

Since adding salt to the water caused the carrot to float, the density of the water must have been increasing.

- What can you conclude about the weight of a carrot slice compared to the weight of an equal volume of fresh water? Salt water?

Since the carrot sinks in fresh water, the carrot must be more dense than fresh water. Therefore, a carrot slice must weigh more than an equal volume of fresh water. Since the carrot floats in salt water, it must be less dense than salt water. Therefore, a carrot slice must weigh less than an equal amount of salt water.

5. Show students a carrot slice "hovering" in a cup of salt water and challenge them to make their own carrot slice hover.

Teacher preparation: Prepare the "hovering" carrot slice ahead of time by following the procedure below. Show students the hovering carrot slice and the materials that you used to make it hover but don't reveal your method to students. Then give students the chance to try to make their own carrot slice hover.

Procedure

1. Half-fill a tall clear plastic cup with room-temperature water.
2. Place a thin slice of carrot in the cup. It should sink because it is more dense than water.

3. Add about 1 heaping teaspoon of salt and stir with a spoon until as much salt dissolves as possible.
4. Continue adding salt and stirring until the carrot floats to the top.
5. Very carefully add fresh water to the top of the salt water until the carrot begins to sink.
6. If the carrot sinks to the bottom, add small amounts of salt and fresh water as needed to cause it to hover.

6. Have students write how their group was able to make the carrot slice "hover" or float in the middle of the cup.

Pass out *Activity sheet 6.3—Changing the density of a liquid* so that students can record their group's procedure and explain how the density of the carrot compared with the density of water and salt water.

7. Have students share their procedures and results.

Students may have used different methods to get their carrot slices to hover, but the method used should be similar to the procedure described in Step 5. The carrot slice will hover when the density of the salty water and the density of the carrot slice are about the same.

Ask students what effects whether something will sink or float in a liquid. Students should realize that whether or not something sinks or floats in a liquid will depend on its density compared to the density of the liquid.

Activity 6.4
Changing the density of an object

Question to investigate

Can changing the shape of an object affect whether it sinks or floats?

Throughout the activities in this investigation, students may have wondered how a boat made out of steel, which is more dense than water, could float. This activity addresses that question. Students will see that changing the shape of an object, like a clay ball, that is more dense than water, can affect whether the object will sink or float. The density of the clay used in this activity does not change, but the density of the object made from the clay does.

1. Discuss with students how to make clay float.

Remind students that a lump of clay sinks. Tell them that most metal, like steel, sinks but that many big boats are made of out of steel. Then ask students how they might be able to make a lump of clay float.

2. As a demonstration, make a clay box that will sink.

If someone suggests making a boat, bowl, or box shape, form the ball of clay into a small, thick-sided box. Place the clay box into a clear plastic cup of water to show that it sinks.

Teacher note: This demonstration uses a box shape so that you can easily calculate the volume of the object. At the end of the activity, you can make a box with a larger volume and show students that this increase in volume decreased the density enough so that the clay could float. At the end of the activity, if you show students the formula for density, d = m/v, they can see that increasing the volume will decrease the density.

3. Have groups try to make clay float and discuss the results.

Ask students how they could change a ball of clay to make it float. Give each group one ball of clay and allow them a short amount of time to try to get it to float. Compare the clay objects students make and how well they float. Ask students what features seem to make a piece of clay float best.

4. Have students make two clay boxes of different volumes.

Show students the same small thick-sided clay box that they saw sink in water in the demonstration. Ask students why they think this clay box sinks and what can be done to make it float. Then have students try to make a clay box that floats. The following procedure is an example of one that the students might try as they attempt to make a clay box that floats.

Procedure

1. Flatten one ball of clay into a large thin pancake shape about 10 cm or more in diameter.

2. Bend the edges up on the clay pancake to make a large shallow open box.

3. Add water to a cup until it is about ¾ full.

4. Slowly and carefully place your clay box on the surface of the water. It should float. If it does not float, remove the clay box from the water and try increasing its volume again.

5. Discuss student observations.

Expected results: **Students' larger thin-sided boxes should float.** Ask students what made their clay box float better than the one you made. Students should realize that the larger volume of their boxes allowed them to float. Point out to students that increasing the volume of the clay box made the boxes less dense so that they could float. Tell students that the density of the clay itself didn't change, but the density of the clay boxes did. Tell them that as the volume increases, the density of the clay box decreases. At some point, the clay box becomes less dense than water and will float.

Teacher note: If you choose, you may model how to calculate the volume of the clay boxes by measuring (in cm) the length, width, and height of your small thick-sided box. Students could then measure and calculate the volume of their larger clay boxes. You could weigh (in grams) the small thick-sided clay box that you made and calculate its density using $d = m/v$. Then have students do the same with their clay boxes. The density of the students' boxes should be less than yours. Theirs should be less than the density of water, which is 1 g/cm^3, while yours should be greater than the density of water.

Ask students why they think a heavy ship made out of steel can float. They should be able to say that although the density of the steel is greater than water, the steel is shaped into such a large volume that the density of the ship is less than the density of water, so it floats.

Demonstration 6b
Applying density to sink and float

Question to investigate

Why does a small pebble sink in water, while a heavier piece of wood floats?

In *Demonstration 6a—Introducing density through sink and float*, students saw that heavy things do not always sink and light things do not always float. As they conducted the activities in this investigation, students discovered that the density of the object as well as the density of the liquid, determines whether or not an object will float or sink. This demonstration and its corresponding Demonstration sheet serve as an assessment. Students should now have the vocabulary and understanding to explain why the small pebble sinks in water while the heavier piece of wood floats.

1. Demonstrate that the wood is heavier than the pebble, yet floats in water.

- Place the pebble on one end of a balance scale.
- Place the block of wood on the other end so that it is obvious that the wood is much heavier than the pebble.
- Then place the pebble in a clear container of water and let it sink.
- Place the wood in a clear container of water and let it float.

2. Have students explain, in terms of density, why the heavy wood floats and the lighter pebble sinks.

Pass out *Demonstration sheet 6b—Applying density to sink and float* and have students explain this phenomenon in writing. Students should explain that the reason any object sinks or floats in water has to do with how the object's density compares with the density of water. If the object is less dense than water, it will float. If it is more dense, it will sink. The piece of wood in this demonstration must weigh less than an equal volume of water, so it is less dense than water and floats. The pebble must weigh more than an equal volume of water, so it is more dense than water and sinks.

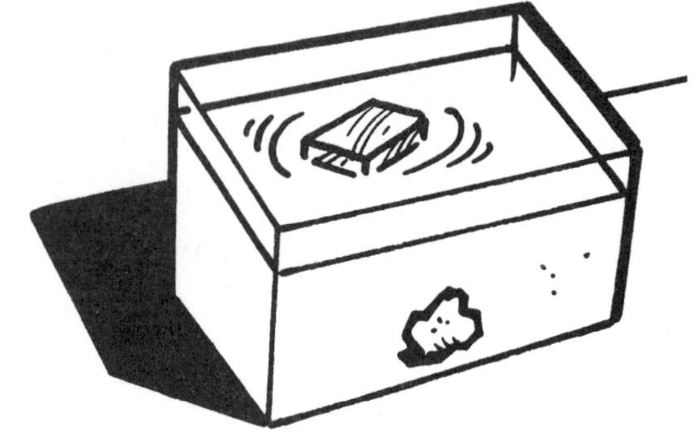

Name:______________________________

Activity sheet 6.1

Comparing the density of an object to the density of water

Place a check in the boxes below to show what you observed when you compared equal volumes of the following substances.

	water	wax	clay
Which is the lightest?			
Which floats in water?			
Which is the heaviest?			
Which sinks in water?			

Fill in the blank with *float* or *sink*:

If a substance weighs less than an equal volume of water, it will __________________.
If a substance weighs more than an equal volume of water, it will __________________.

Fill in the blank with *more dense* or *less dense*:

If a substance weighs less than an equal volume of water, it is ____________________ than water.
If a substance weighs more than an equal volume of water, it is ____________________ than water.

Draw a picture of a piece of clay in a cup of water.

Draw a picture of a piece of wax in a cup of water.

Rank water, clay, and wax according to their densities.

least dense________________

____

most dense________________

An ice cube will float in a cup of water. What would you expect if you compared the weight of the ice cube to the weight of an *equal volume* of liquid water?

____

____

Name:______________________________

Activity sheet 6.2

Comparing the density of different liquids

When comparing equal volumes of the following liquids, which liquid is the heaviest, lightest, and in between?

Water ______________________

Vegetable oil ______________________

Corn syrup ______________________

What did you do to make sure that you compared equal volumes of water, vegetable oil, and corn syrup?

____

____

Since you used equal volumes of the liquids, you can compare the weight of each to find out about the density of each liquid. List the liquids in order from the least dense to the most dense.

Least dense ______________________

____

Most dense ______________________

Draw your observations of the liquids and objects in the cup and label them.	Use what you know about density to explain why the liquids and objects are positioned where they are in the cup.

Name:______________________________

Activity sheet 6.3

Changing the density of a liquid

1. Write what your group did to make the carrot float in the middle of the cup.

____

____

____

____

____

____

____

____

2a. Before you added salt, did the carrot sink or float in the water? __________________

2b. How did the density of the carrot compare with the density of the water?

____

____

3a. After you added salt, the carrot began to __________________________.

3b. How did the density of this salt water compare with the density of the water before you added any salt?

____

____

Name:______________________________

Demonstration sheet 6b

Applying density to sink and float

Use what you have learned about density in this investigation to answer the following question.

Why does a lighter pebble sink in water, while a heavier piece of wood floats?

____

____

____

____

____

____

____

____

____

____

Name:______________________________

Assessment rubric

Investigation 6—Density

To earn a "B," a student must receive a "Very good" in each category.

	Very good	Satisfactory	Needs improvement
Activity sheet 6.1	________	________	________
Activity sheet 6.2	________	________	________
Activity sheet 6.3	________	________	________
Demonstration sheet 6b	________	________	________
Investigative behaviors	________	________	________

Activity sheet 6.1

Comparing the density of an object to the density of water

Compares relative densities of wax and clay to water
Uses evidence from observations to help form explanations
Applies learning to hypothetical situation

Activity sheet 6.2

Comparing the density of different liquids

Weighs liquids accurately
Makes labeled drawing that reflects observations
Explains the relative densities of the liquids and solids

Activity sheet 6.3

Changing the density of a liquid

Writes detailed procedure that can be followed
Identifies relative density of carrot, salt water, and water to explain why the carrot sinks or floats

Demonstration sheet 6b

Applying density to sink and float

Realizes density of water is factor in sinking and floating
Refers to mass and volume in explanation

Investigative behaviors

Uses a balance scale to weigh substances

Participates in design of experiments
Participates in class discussions
Works cooperatively with group
Uses evidence to formulate explanations

To earn an "A," a student must also exhibit some of the following qualities throughout this investigation.

Shares detailed observations
Participates well in class discussions
Participates well in group work
Uses scientific thinking
Consistently exhibits exceptional thought and effort in tasks

Investigation 7. Mixtures and solutions

Summary

Students will add solids to various liquids and observe whether and how well they break apart in the liquids. Students will realize that there are different types of mixtures depending upon how completely a substance is mixed within another. They will also add three liquids to water to discover that mixing and dissolving apply to liquids as well as solids. Students will also investigate a gas dissolved in a liquid. These experiences will help students develop a meaning of *dissolving* and recognize that *solubility*, that is, whether and how well a substance breaks apart or dissolves is a characteristic property of a substance.

Objective

Students will develop a definition of mixture, dissolving, and solution; help design an investigation to explore dissolving; and recognize that dissolving involves solids, liquids, and gases.

Assessment

The assessment rubric *Mixtures and Solutions* on page 179 is included so that you can assess and document student progress throughout the investigation. The abilities and understandings demonstrated by students and recorded on the rubric include the following: As students add solids to different liquids and see the different mixtures that result, they will begin to develop a definition of "dissolve." They will also see that liquids and gases can also dissolve in liquids. Students will add color, sugar, and chocolate to three different liquids and use the results to predict how a crushed M&M will mix in the liquids. Investigative behaviors observed as students help design and conduct the activities, work cooperatively with their group, and develop explanations based on their results can also be recorded on this scoring rubric.

In the final activity, students will be challenged to keep as much dissolved gas as possible in a lemon soda that they make. This challenge may be used as a performance assessment. In order for students to successfully complete this activity, they must demonstrate mastery of certain concepts. By observing students as they attempt to make lemon soda that retains much of its carbonation, you can informally evaluate to what extent students have mastered the abilities, understandings, and investigative behaviors developed in the investigation.

Relevant *National Science Education Standards*

Physical Science

K–4

Properties of Objects and Materials

Objects have many observable properties.

5–8

Properties and Changes of Properties in Matter

A substance has characteristic properties such as density...and solubility, all of which are independent of the amount of the sample.

Science as Inquiry

K–4

Abilities Necessary to do Scientific Inquiry

Ask a question about objects.
Plan and conduct a simple investigation.
Employ simple equipment and tools to gather and extend the senses.
Use data to construct a reasonable explanation.
Communicate investigations and explanations.

Understandings about Scientific Inquiry

Scientific investigations involve asking and answering a question.
Scientists use different kinds of investigations depending on the questions they are trying to answer.
Types of investigations include describing objects...and doing a fair test.
Good explanations are based on evidence from investigations.

5–8

Abilities Necessary to do Scientific Inquiry

Identify questions that can be answered through scientific investigations.
Design and conduct a scientific investigation.
Use appropriate tools and techniques to gather, analyze, and interpret data.
Develop descriptions, explanations, predictions, and models using evidence.
Think critically and logically to make the relationships between evidence and explanations.
Communicate scientific procedures and explanations.

Understandings about Scientific Inquiry

Different kinds of questions suggest different kinds of scientific investigations.
Scientific explanations emphasize evidence and have logically consistent arguments.
Scientific investigations sometimes result in new ideas and phenomena for study or generate new procedures for an investigation. These can lead to new investigations.

How this investigation relates to the *Standards*

One goal of the *Standards* is for students to understand that substances have characteristic properties that distinguish them from other substances. Whether and to what extent a substance dissolves in a particular liquid is a characteristic property of the substance. In these activities, students will design and conduct investigations to explore how solids, liquids, and gases dissolve in various liquids. The results of these activities will provide evidence that solubility is a characteristic property of a substance.

Materials chart

7.1 Different kinds of mixtures

7.2 Solids mix differently in different liquids

7.3 Gases can dissolve in liquids

Each group will need	Activities 7.1	7.2	7.3
water in source cup		•	
club soda in cup, ≈ 1 cup			•
grape Kool-Aid in cup	•		
powdered baking cocoa in cup	•		
white paper, 8½ x 11		•	
clear plastic cups	2	9	3
Popsicle sticks	2	9	
masking tape		•	
small cup, 3½ ounces			1
colored sugar in cup, ≈ 1 tablespoon		•	
sugar in cup, ≈ 1½ tablespoons			•
piece of uncooked pasta			•
pipe cleaner			•
M&Ms, same color		3	≥ 3
chocolate chips		3	
isopropyl alcohol in cup		•	
vegetable oil in cup		•	
corn syrup in cup		•	
lemon juice			•
plastic teaspoon		•	•
paper towel		•	•
magnifying glass			•

Notes about the materials

When using isopropyl alcohol, read and follow all warnings on the label. Be sure students are wearing properly fitting goggles.

Teacher will need...

to prepare for the students

7.2 Solids mix differently in different liquids

for the demonstrations

Demonstration 7a—Liquids mix differently in water

7.3 Gases can dissolve in liquids (page 173, step 2)

	Activity 7. 2	Demo 7a	Activity 7.3
sugar	•		
zip-closing bag	•		
tablespoon measure	•		
food coloring	•		
water		•	
isopropyl alcohol		•	
vegetable oil		•	
corn syrup		•	
clear plastic cups		3	
spoons or Popsicle sticks or straws		3	
unopened bottle of club soda			•

Notes about the materials

You will need about 2 liters of club soda for 8 groups of students to do activity 7.3. You may use this same 2-liter bottle in the demonstration in activity 7.3. Remove the label before the demonstration.

Teacher preparation for activity 7.2

1. Prepare colored sugar by placing 1 tablespoon of sugar, times the number of science groups in your class, into a sandwich-sized zip-closing plastic bag.

2. Then add 2 drops of food coloring for each tablespoon of sugar. The food coloring should be the same color as the M&Ms used in the activity. Red works well.

3. Seal the bag, leaving as much air in as you can. Shake the bag vigorously until the sugar is thoroughly colored.

Activity sheets

Copy the following activity sheets or provide blank paper and distribute them as specified in the activities.

Activity sheet 7.2 page 176
Solids mix differently in different liquids

Copy one per student.

Students will record their observations on this activity sheet.

Demonstration sheet 7a page 177
Liquids mix differently in water

Copy one per student.

Students will record what they observe in the demonstration and determine if each of the liquids dissolves in water. Students will write a definition for "dissolved."

Activity sheet 7.3 page 178
Gases can dissolve in liquids

Copy one page for every two students.

Cut the page in half so that each student will receive his or her own set of directions. Also provide one piece of white construction paper to each student. Students will fold the construction paper in half like a greeting card, title each "page," and respond to the prompts from this activity sheet, as they describe their group's procedure for making and testing their fizzy lemon soda.

Science background information

A mixture is created when two or more types of matter are combined but do not undergo a chemical reaction. The components of a mixture may stay distinct and easily distinguished like a mixture of salt and pepper. Or the components may mix very thoroughly so that they cannot be recognized as separate types of matter like a mixture of salt in seawater.

Whether, and to what extent, a substance spreads out and mixes within another substance is called its *solubility*. Solubility is a characteristic property of a substance. When talking about solubility, the smaller amount of substance being added is called the *solute*. The substance the solute is placed in is called the *solvent*. If the solute mixes and spreads completely throughout the solvent so that the particles are so intimately intermingled with those of the solvent that they will not settle out, then the substance has *dissolved* in the solvent. The combination of dissolved solute and solvent is called a *solution*. This can be distinguished from a *suspension* in which the solute will eventually settle out.

When students think about dissolving, they usually think about solids such as salt, sugar, or drink mix dissolving in a liquid—usually water. But dissolving can apply to all states of matter. A gas can dissolve in a gas, such as in the mixture of gases in air, or a gas can dissolve in a liquid, such as the carbon dioxide gas dissolved in water to make soda pop. A solid can even be dissolved in a solid, such as tin dissolved in copper to make bronze.

The activities in this investigation will only deal with solids, liquids, and gases dissolving in liquids. Students will realize that even though a solid dissolves in water, it will not necessarily dissolve in other liquids. They will also see that even though a particular liquid dissolves in water, other liquids will not necessarily do the same.

Whether or not a substance dissolves in a liquid has to do with the way the particles of the solute and solvent interact on the molecular level. Water is an excellent solvent for many substances like salt and sugar, where the positive and negative charges on the particles that compose the substances can interact with the positive and negative areas of the polar water molecules. Water is not an effective solvent for non-polar substances such as wax or oil.

Activity 7.1—Different kinds of mixtures

The Kool-Aid dissolves very well in water, and the cocoa does not dissolve well at all. One of the main ingredients in Kool-Aid is citric acid. The molecular structure of citric acid makes it very soluble in water. It has lots of areas on it where a hydrogen and an oxygen are bonded together, causing a negatively charged area near the oxygen and a positively charged area near the hydrogen. These polar regions interact with the positive and negative areas of the water molecules, making citric acid extremely soluble.

The cocoa is composed of molecules called *fatty acids*. These molecules have a very low proportion of charged areas and are not very soluble in water.

Activity 7.2—Solids mix differently in different liquids

Testing chocolate in each liquid

The chocolate breaks apart to various degrees in the different liquids. Chocolate is made from cocoa which is composed of a type of molecule called a *fatty acid.* Fatty acids are long molecules that are relatively non-polar. They are not very soluble in water. The differences in the extent to which the chocolate breaks down in the different liquids is mostly attributable to the extent to which the fatty acids are soluble in these different liquids.

Testing colored sugar in each liquid

Sugar and food coloring are very soluble in water. Sugar has several places on it that are polar and interact readily with water molecules. Both the coloring and the sugar should dissolve pretty well in water. Alcohol has some polar characteristics but not as much as water. The food coloring dissolves into the alcohol but the sugar does not dissolve in the alcohol as much as it does in the water. Oil molecules are non-polar. Neither the coloring or the sugar dissolved well in oil.

Testing an M&M in each liquid

Since an M&M is, to a large extent, sugar, coloring, and chocolate, the results of the previous two tests should help students predict the extent to which these components of an M&M will dissolve in the different liquids.

Demonstration 7a—Liquids mix differently in water

Most of us don't think about liquids dissolving in liquids but the same concepts apply as solids dissolving in liquids. In both cases, the molecules of the added substance (solute) are struck by and become surrounded by molecules of the solvent. Depending on the interactions and degree of attraction between the molecules of solute and solvent, the solvent may attach to the solute and break it apart to a greater or lesser extent. If the molecules of the solute (whether liquid or solid) become so intermingled and closely associated with the molecules of the solvent that they do not separate, then dissolving has occurred. The degree to which the three different liquids dissolve in water is a function of the degree to which the polar water molecules interact with the molecules of the other liquids.

Activity 7.3—Gases can dissolve in liquids

In a soda pop factory, carbon dioxide gas is injected, at high pressure, into each container of cold soda, and the container is then sealed. The carbon dioxide gas becomes thoroughly mixed and dissolved throughout the liquid soda to become part of the soda solution. When students feel the outside of the soda bottle, it feels very hard because of the pressure pushing out from the inside of the bottle. When the pressure is released, the gas comes out of the solution and bubbles up out of the soda.

When students place a noodle, pipe cleaner, and sugar into the soda, they see bubbles collect on these objects and substances. The gas that is still in the solution collects on the surfaces of these objects and forms bubbles, which rise in the liquid. Bubbles form and rise from the M&M for the same reason. The surface of the M&M may seem smooth but it is rough enough on the microscopic and molecular level to have many points where the gas molecules can attach, collect, and form bubbles.

When sugar is added to the club soda to make a lemon soda, the many surfaces of the sugar granules provide numerous places where the carbon dioxide gas molecules can attach and collect to form bubbles. Adding sugar directly to the soda takes gas out of the soda, causing it to become "flat." But dissolving the sugar in the lemon juice *before* adding it to the soda significantly reduces the number of rough surfaces and reduces the ability of the gas to collect and bubble out of the soda. This helps the soda keep its fizz.

Activity 7.1
Different kinds of mixtures

Question to investigate

How can you tell a solution from a different kind of mixture?

In this activity, students will compare the way Kool-Aid mixes with water to the way cocoa mixes with water. Students will not see any particles in the transparent and colored Kool-Aid mixture, but will see many in the cocoa mixture. Through these observations, students will begin to develop an understanding of the characteristics of a *solution* and that whether or not a substance *dissolves* in a liquid is a property of the substance and the liquid it is placed in.

1. Have students add Kool-Aid to water and cocoa to water.

Procedure

1. Half-fill 2 clear plastic cups with room-temperature water.
2. Use a Popsicle stick to sprinkle a small amount of Kool-Aid onto the surface of one of the cups of water.
3. Sprinkle a similar amount of cocoa onto the surface of the other cup of water. Observe both cups from the side.
4. Stir the powder in each cup.

2. Discuss student observations.

Ask students what they observed when they first placed each powder onto the surface of the water. Then ask them what they observed when they stirred each powder.

Expected results: **The two powders mix into water differently. The Kool-Aid combines very quickly when sprinkled on the surface of the water. When stirred, it dissolves into the water so that the water is completely colored and no particles are visible. The cocoa slowly drops from the surface into the water. When stirred, it colors the water but has visible particles in it.**

Tell students that the Kool-Aid is dissolved in the water and that the cocoa broke into small particles but did not dissolve. Ask students what they should look for to see if a substance dissolves in a liquid. Students should conclude that if a substance mixes in so thoroughly that no particles are visible after stirring and every part of the mixture looks the same, then the substance has dissolved. Tell students that when a substance is dissolved in another, the mixture of the substances is called a solution. You could also tell students that if the Kool-Aid and water truly made a solution, the Kool-Aid should stay in the solution and not settle out.

Activity 7.2
Solids mix differently in different liquids

Question to investigate

How can you predict whether an M&M will dissolve in different liquids?

In this activity, students will attempt to dissolve chocolate and colored sugar in water, alcohol, and oil. Students will see that these substances break apart differently depending on the liquid they are placed in. Since chocolate and colored sugar are ingredients in M&Ms, students can use their results to predict how the colored candy coating and the chocolate of an M&M will break apart when placed in these same liquids.

1. Introduce the activity.

Tell students that water, alcohol, and oil are all commonly used to dissolve different substances. Depending on what the substance is, it may dissolve better in one of these liquids than another. For example, vanilla can be dissolved in a type of alcohol to make vanilla extract, but does not dissolve well in water. Some vitamins, like Vitamin C, dissolve best in water while others, like Vitamin E, dissolve best in oil. One of the properties of a substance is how well it dissolves in a particular liquid.

Tell students that in this activity, they will see to what extent colored sugar and chocolate dissolves in three different liquids. Tell them that since an M&M is basically made of a colored sugar coating and milk chocolate, they can use their results to predict how an M&M will dissolve in each of the three liquids.

2. Discuss with students how to set up an experiment to compare how colored sugar mixes in water, alcohol, and vegetable oil.

Ask students how they would set up an investigation to see how well colored sugar mixes in water, alcohol, and oil. Students should suggest using the same amount of the same colored sugar in the same amount of each liquid. They should also suggest stirring each in the same way and for the same amount of time.

3. Have students mix colored sugar in water, alcohol, and oil.

Procedure

1. Use masking tape and a pen to label 3 clear plastic cups *water*, *alcohol*, and *oil*. Place 1 teaspoon of water, alcohol, and vegetable oil into the labeled cups.

2. Add ½ teaspoon of the colored sugar to each of the cups and stir each with a clean Popsicle stick or straw.

Have students record their observations on *Activity sheet 7.2—Solids mix differently in different liquids* on page 176.

4. Discuss student observations.

Ask students questions like the following:

- What do you observe in each cup?
- Does the color seem to dissolve more in one liquid than in another?
- Does the sugar seem to dissolve more in one liquid than in another?

Expected results:
***Colored sugar in water*—The color and the sugar dissolve completely in the water.**
***Colored sugar in alcohol*—The color dissolves, but the sugar does not dissolve.**
***Colored sugar in oil*—The color does not dissolve, and neither does the sugar.**

5. Discuss with students how to set up an experiment to compare how milk chocolate mixes in water, alcohol, and vegetable oil.

Ask students what they would need to do in order to compare the way chocolate mixes with each of the liquids. Students should suggest using the same amount of chocolate in the same amount of each liquid. They should also stir each in the same way and for the same amount of time. Tell students that they will also need to crush the chocolate chips so that the liquid touches as much chocolate as possible.

6. Have students mix a crushed chocolate chip into water, alcohol, and oil.

You can use this procedure or any similar one that your class designs.

Procedure

1. Use masking tape and a pen to label 3 clear plastic cups *water*, *alcohol*, and *oil*.

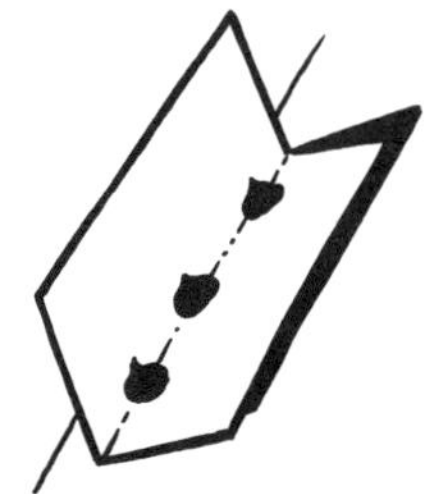

2. Fold a piece of paper in half lengthwise and place 3 chocolate chips along the inside fold about an equal distance apart. Close the paper over the chocolate chips.
3. Press down on the paper over each chocolate chip with a Popsicle stick or pencil until each is crushed well. Rub each smashed chip gently in your fingers to get it to crumble into tiny pieces.

4. Place 1 teaspoon of water, alcohol, and vegetable oil into the labeled cups. Use a paper towel to wipe the spoon between each liquid.
5. Add a crushed chocolate chip to each liquid. Stir the contents of each cup with a clean Popsicle stick.

Have students record their observations on *Activity sheet 7.2—Solids mix differently in different liquids* on page 176.

7. Discuss student observations.

Ask students questions like the following:

- What do you observe in each cup?
- Does the chocolate seem to break apart more in one liquid than in another?

Expected results:
While the chocolate does not dissolve in any of the liquids, it should cause each liquid to look different.
Chocolate in water— **The chocolate does not break apart much in the water.**
Chocolate in alcohol—**The chocolate does break apart somewhat in the alcohol. Tiny particles are visible.**
Chocolate in oil—**The chocolate breaks apart somewhat in the oil, and tiny particles are visible. It seems to color the oil slightly more than the alcohol.**

8. Have students predict how an M&M will mix in water, alcohol, and oil.

Have students compare how the colored sugar and chocolate mixed in water to the way they mixed in alcohol and oil. Remind students that colored sugar and chocolate are the main components of an M&M. Based on the results of their tests, ask students to predict how each part of a crushed M&M will mix in each liquid. Students should predict that the M&M's color, white sugar coating, and chocolate will mix in each liquid the way the coloring, sugar, and chocolate did in their tests.

9. Have students design and conduct an experiment to test their prediction.

Ask students what they will need to keep the same in order to make their test fair. Students should suggest using the same color M&M in the same amount of each liquid. They should also stir each in the same way and for the same amount of time. Tell students that they will also need to crush the M&M so that the liquid touches as much of the M&M as possible. You can use the following procedure or any similar one that your class designs.

Procedure

1. Use masking tape and a pen to label 3 clear plastic cups *water*, *alcohol*, and *oil*. Place 1 teaspoon of water, alcohol, and vegetable oil into the labeled cups.

2. Fold a piece of paper in half lengthwise and place 3 M&Ms along the inside fold about an equal distance apart. Close the paper over the M&Ms.

3. Press down on the paper over each M&M with a Popsicle stick or pencil until each M&M is crushed well.

4. Add 1 crushed M&M to each liquid and stir with a clean Popsicle stick or straw.

Have students record their observations on *Activity sheet 7.2—Solids mix differently in different liquids* on page 176.

10. Have students share their observations and the accuracy of their predictions.

Ask students what they observed in each cup.

Expected results:
M&M in water—**The color and sugar coating both dissolve, but the chocolate does not break up much at all.**
M&M in alcohol—**The color comes off, but the sugar coating does not dissolve. The chocolate breaks apart somewhat and colors the alcohol.**
M&M in oil—**The color does not come off, and the sugar coating does not dissolve, but the chocolate does break apart a bit and colors the oil.**

Demonstration 7a
Liquids mix differently in water

Question to investigate

Do all liquids mix the same way in water?

Just as solids break apart differently in different liquids, liquids also interact with other liquids in a characteristic way. In this demonstration, students will see three different liquids being poured into water and then stirred. They will observe whether and to what extent these liquids dissolve.

1. Do a demonstration to show students that alcohol, corn syrup, and vegetable oil mix differently in water.

Remind students that the *solids*, Kool-Aid and cocoa, mix differently in water. Ask them if they think different *liquids* will mix into water differently. This demonstration works best if students are close enough to see the contents of the cups clearly.

Procedure

1. Fill three clear plastic cups about ⅔ full with room-temperature water.
2. Slowly pour about 2 tablespoons of corn syrup into the first cup of water.
3. After the corn syrup settles, ask students if the corn syrup is dissolved in the water.

Expected results: **The corn syrup will sink to the bottom and stay there.** It has not yet dissolved.

4. Tell students that sometimes stirring helps substances dissolve. Stir the corn syrup and water until the corn syrup is so thoroughly mixed into the water that it turns clear. Students should realize that the corn syrup is dissolved.
5. In the second cup of water, pour about 2 tablespoons of rubbing alcohol. Ask students if the alcohol has dissolved.

Expected results: **The alcohol is visible as it mixes into the water, but quickly dissolves and turns clear.** There is no need to stir.

6. In the third cup of water, pour about 2 tablespoons of vegetable oil. Ask students if the oil has dissolved.

Expected results: **The oil will form a layer on the surface of the water.**

7. Stir the oil and water.

Expected results: **The oil and water will not mix, and the oil will form a layer on the surface again.**

2. Discuss whether or not all solids and liquids dissolve in water.

Ask students questions like the following:

- Do all liquids dissolve in water?
- Do all liquids mix the same way in water?

Students should realize that not all liquids dissolve in water. Tell them that the way a liquid mixes in water is a characteristic property of the liquid. So far they have seen how solids and liquids mix in water. In the next activity, students will see if gases dissolve in water.

3. Have students write their observations and thoughts about dissolving.

Pass out *Demonstration sheet 7a—Liquids mix differently in water* on page 177.

Activity 7.3
Gases can dissolve in liquids

Question to investigate

How can you make the gas in soda pop come out of or stay in the soda?

The idea of a gas being dissolved in a liquid may be difficult for students to accept. In the same way that some solids and liquids can be dissolved in water, gases can also dissolve in water. Since the carbon dioxide gas in carbonated water is so thoroughly mixed and intermingled throughout the water, the gas is *dissolved* in the water. In this activity, students will use different methods to get carbon dioxide gas to come out of the carbonated water. They will then be challenged to make a lemon soda that retains much of its carbonation.

1. Discuss with students some of their experiences with the bubbles in soda.

Ask students questions like the following:

- Have you ever noticed the bubbles that appear on the inside of a glass of soda or that form on a straw that is placed in soda? Sometimes so many bubbles will form on the straw that the straw will rise up a little bit.
- Where do you think these bubbles come from?

2. Show students the bubbles that appear when you open a new bottle of soda.

Before opening the bottle, ask students if they see any bubbles. Then ask them what they think will happen when you open the bottle of soda. Very slowly unscrew the bottle cap.

Expected results: **Many bubbles will appear throughout the soda.**

Close the bottle again. Tell students that a gas called *carbon dioxide* was dissolved into the soda at the soda factory. Opening the bottle allows some of that gas to come out. Ask students if they think that there is any more gas dissolved in the soda. Students will probably guess, based on their experience drinking sodas, that there is still gas left in the soda.

3. Have students experiment with ways to get dissolved gas out of the soda.

Tell students that in the same way dissolved gas collects on a straw, it can also collect on other objects. Ask students what they could do to detect any gas that might still be left in the soda. Students should suggest placing a straw or other object in the soda.

The following procedure suggests some ways to remove dissolved gas from soda. You and your students may think of other objects to place in soda that will also remove some of the dissolved carbon dioxide.

Procedure

1. Fill a clear plastic cup about ¼ of the way with club soda and replace the cap on the bottle. While watching from the side, place ½ of an uncooked rotini noodle into the cup of soda. Use a magnifying glass to get a better look.

2. Bend a pipe cleaner to make a hook and use it to take the noodle out of the soda. Watch the surface of the pipe cleaner while it is in the soda.

3. Sprinkle a little sugar onto the surface of the soda and observe.

4. Place an M&M in the soda. Watch it closely.

4. Introduce students to the problem of keeping the carbonation in soda.

Ask students what they observe after the addition of each object.

Expected results: **Bubbles will appear on the noodle, pipe cleaner, sugar crystals, and M&M. This happens because carbon dioxide gas, which was dissolved in the water, adheres and collects on the bumps and rough spots on the objects to form bubbles, which float to the surface.**

Tell students that it seems that no matter what they add to the carbonated water, some of the dissolved gas comes out. If enough dissolved gas escapes, the soda is no longer fizzy. Ask students how they think soda factories add flavoring and sweetener to the carbonated water without letting all of the gas escape. Tell students that they will do an experiment to find a way to add flavoring and sweetener to carbonated water so that it doesn't lose too much carbon dioxide gas.

5. Give students the recipe for a lemon soda and have them make it.

Tell students to make a lemon soda according to the following procedure.

Procedure

1. Fill your clear plastic cup about ¼ of the way with club soda.

2. Add 1 teaspoon of lemon juice. Stir with the spoon until the lemon juice is completely dissolved.

3. Add 1 teaspoon of sugar. Stir until the sugar is completely dissolved.

6. Have students report their observations.

Have students tell what they observed when the lemon juice was added and then when the sugar was added.

Expected results: **Not many bubbles come out of the soda when lemon juice is added, but many bubbles foam up when sugar is added and then stirred.**

7. Have students test their sodas for carbonation.

Tell students that they need to test the amount of carbonation left in the soda by placing an M&M in it.

They will later compare the amount of bubbling from the surface of this M&M to the amount of bubbling from an M&M placed in another lemon soda that they will make.

Procedure

1. Place an M&M in the soda. Observe the amount of bubbling from the surface of the M&M. This will give you some idea of the amount of carbon dioxide gas left in the soda. Remember what this looks like.

Expected results: **A thin stream of tiny bubbles rises slowly from the M&M. This shows that there is some carbonation left in the soda.**

8. Challenge students to develop a better method of making a lemon soda so that less gas is lost in the process.

Tell students to develop a different method for making a lemon soda that keeps its fizz better than the lemon soda they just made. They should use the same amount of lemon juice, sugar, and carbonated water. They should also use an M&M to compare the amount of carbon dioxide gas left in the soda using their new method with the amount left from the first method. The following procedure is one way to accomplish this.

Procedure

1. In a separate cup, combine 1 teaspoon sugar and 1 teaspoon lemon juice. Stir until the sugar dissolves.

2. Fill your clear plastic cup about ¼ of the way with club soda.

3. Pour the sugar and lemon juice solution into the carbonated water and stir until it dissolves.

4. Place an M&M in the soda. Compare the amount of bubbling from this M&M to the amount of bubbling from the M&M in the first method.

9. Discuss student results and apply their learning to a real-life example.

Have students share the methods they devised to make a fizzier lemon soda.

Expected results: **Many bubbles rise in a stream from the M&M. These bubbles seem to be larger and in greater quantity than the bubbles that rose from the M&M in the first lemon soda.** Students can conclude that their method keeps more carbon dioxide gas in the soda.

Tell students that syrups are often used to flavor sodas. If they look at the ingredient list on a can of soda pop, they will see that corn syrup is used as a sweetener. Have them explain at least one reason why they think that corn syrup is used instead of granulated sugar.

10. Have students write about their fizzy lemon soda.

Pass out the directions found on *Activity sheet 7.3—Gases can dissolve in liquids* on page 178. Also give students a white piece of construction paper to write about their group's procedure for making a fizzy lemon soda.

Name:______________________________

Activity sheet 7.2

Solids mix differently in different liquids

What do these solids look like when they are mixed with these liquids?

		water	alcohol	oil
colored sugar	coloring			
	sugar			
chocolate	chocolate			
M&M	coloring			
	sugar coating			
	chocolate			

Name:______________________________

Demonstration sheet 7a

Liquids mix differently in water

When ___ is poured into water	What do you observe?	Does the liquid dissolve in water?
alcohol		
oil		
corn syrup		

How can you tell if a liquid is dissolved in water?

____

____

How can you tell if a liquid is not dissolved in water?

____

____

What does it mean if something is dissolved in something else?

____

____

____

Activity sheet 7.3

Gases can dissolve in liquids

Fold a piece of white construction paper in half like a greeting card. Then label each page with the following titles. You will then write about your experiment on each page.

On the outside front cover
Write a title for your mini-report like: "How to Make a Fizzier Lemon Soda" or "The Secret to Making a Fizzier Lemonade." You may title this mini-report as you see fit. Be sure to draw a picture of a fizzy lemon soda, too.

On the inside left page Title: "Problems to Overcome"
List the problems your group experienced as you made the lemon soda with the first method.

On the inside right page Title: "A Better Way"
Describe your group's most successful method for making a lemon soda that kept its fizz.

On the outside back cover Title: "Testing for Carbonation"
Explain how you used the M&M test to show that your group's method for making lemon soda is better than the first method.

Activity sheet 7.3

Gases can dissolve in liquids

Fold a piece of white construction paper in half like a greeting card. Then label each page with the following titles. You will then write about your experiment on each page.

On the outside front cover
Write a title for your mini-report like: "How to Make a Fizzier Lemon Soda" or "The Secret to Making a Fizzier Lemonade." You may title this mini-report as you see fit. Be sure to draw a picture of a fizzy lemon soda, too.

On the inside left page Title: "Problems to Overcome"
List the problems your group experienced as you made the lemon soda with the first method.

On the inside right page Title: "A Better Way"
Describe your group's most successful method for making a lemon soda that kept its fizz.

On the outside back cover Title: "Testing for Carbonation"
Explain how you used the M&M test to show that your group's method for making lemon soda is better than the first method.

Name:______________________________

Assessment rubric

Investigation 7—Mixtures and solutions

To earn a "B," a student must receive a "Very good" in each category.

	Very good	Satisfactory	Needs improvement
Activity sheet 7.2 Solids mix differently in different liquids	______	______	______
Demonstration sheet 7a Liquids mix differently in water	______	______	______
Activity sheet 7.3 Gases can dissolve in liquids	______	______	______
Investigative behaviors	______	______	______

Activity sheet 7.2
Solids mix differently in different liquids

Records detailed observations in chart
Identifies characteristics of substances

Demonstration sheet 7a
Liquids mix differently in water

Records observations in chart
Develops working definition of "dissolved"

Activity sheet 7.3
Gases can dissolve in liquids

Draws fizzy soda on cover
Labels pages correctly
Lists problems encountered with first soda-making method
Describes successful method for making lemon soda
Explains testing method for fizzier lemon soda

Investigative behaviors

Participates in design of experiments
Participates in class discussions
Works cooperatively with group
Uses evidence to formulate explanations

To earn an "A," a student must also exhibit some of the following qualities throughout this investigation.

Shares detailed observations
Participates well in class discussions
Participates well in group work
Uses scientific thinking

Reprinted with permission from the *National Science Education Standards*, National Research Council, Washington, D.C.: National Academy Press; 1996.

Content Standards: K–4

Science as Inquiry

CONTENT STANDARD A:
As a result of activities in grades K–4, all students should develop
- **Abilities necessary to do scientific inquiry**
- **Understandings about scientific inquiry**

DEVELOPING STUDENT ABILITIES AND UNDERSTANDING

From the earliest grades, students should experience science in a form that engages them in the active construction of ideas and explanations and enhances their opportunities to develop the abilities of doing science. Teaching science as inquiry provides teachers with the opportunity to develop student abilities and to enrich student understanding of science. Students should do science in ways that are within their developmental capabilities. This standard sets forth some abilities of scientific inquiry appropriate for students in grades K–4.

In the early years of school, students can investigate earth materials, organisms, and properties of common objects. Although children develop concepts and vocabulary from such experiences, they also should develop inquiry skills. As students focus on the processes of doing investigations, they develop the ability to ask scientific questions, investigate aspects of the world around them, and use their observations to construct reasonable explanations for the questions posed. Guided by teachers, students continually develop their science knowledge. Students should also learn through the inquiry process how to communicate about their own and their peers' investigations and explanations.

There is logic behind the abilities outlined in the inquiry standard, but a step-by-step sequence or scientific method is not implied. In practice, student questions might arise from previous investigations, planned classroom activities, or questions students ask each other. For instance, if children ask each other how animals are similar and different, an investigation might arise into characteristics of organisms they can observe.

Full inquiry involves asking a simple question, completing an investigation, answering the question, and presenting the results to others. In elementary grades, students begin to develop the physical and intellectual abilities of scientific inquiry. They can design investigations to try things to see what happens—they tend to focus on concrete results of tests and will entertain the idea of a "fair" test (a test in which only one variable at a time is changed). However, children in K–4 have difficulty with experimentation as a process of testing ideas and the logic of using evidence to formulate explanations.

GUIDE TO THE CONTENT STANDARD
Fundamental abilities and concepts that underlie this standard include

ABILITIES NECESSARY TO DO SCIENTIFIC INQUIRY

ASK A QUESTION ABOUT OBJECTS, ORGANISMS, AND EVENTS IN THE ENVIRONMENT. This aspect of the standard emphasizes students asking questions that they can answer with scientific knowledge, combined with their own observations. Students should answer their questions by seeking information from reliable sources of scientific information and from their own observations and investigations.

PLAN AND CONDUCT A SIMPLE INVESTIGATION. In the earliest years, investigations are largely based on systematic observations. As students develop, they may design and conduct simple experiments to answer questions. The idea of a fair test is possible for many students to consider by fourth grade.

EMPLOY SIMPLE EQUIPMENT AND TOOLS TO GATHER DATA AND EXTEND THE SENSES. In early years, students develop simple skills, such as how to observe, measure, cut, connect, switch, turn on and off, pour, hold, tie, and hook. Beginning with simple instruments, students can use rulers to measure the length, height, and depth of objects and materials; thermometers to measure temperature; watches to measure time; beam balances and spring scales to measure weight and force; magnifiers to observe objects and organisms; and microscopes to observe the finer details of plants, animals, rocks, and other materials. Children also develop skills in the use of computers and calculators for conducting investigations.

USE DATA TO CONSTRUCT A REASONABLE EXPLANATION. This aspect of the standard emphasizes the students' thinking as they use data to formulate explanations. Even at the earliest grade levels, students should learn what constitutes evidence and judge the merits or strength of the data and information that will be used to make explanations. After students propose an explanation, they will appeal to the knowledge and evidence they obtained to support their explanations. Students should check their explanations against scientific knowledge, experiences, and observations of others.

COMMUNICATE INVESTIGATIONS AND EXPLANATIONS. Students should begin developing the abilities to communicate, critique, and analyze their work and the work of other students. This communication might be spoken or drawn as well as written.

UNDERSTANDINGS ABOUT SCIENTIFIC INQUIRY

- Scientific investigations involve asking and answering a question and comparing the answer with what scientists already know about the world.
- Scientists use different kinds of investigations depending on the questions they are trying to answer. Types of investigations include describing objects, events, and organisms; classifying them; and doing a fair test (experimenting).
- Simple instruments, such as magnifiers, thermometers, and rulers, provide more information than scientists obtain using only their senses.
- Scientists develop explanations using observations (evidence) and what they already know about the world (scientific knowledge). Good explanations are based on evidence from investigations.
- Scientists make the results of their investigations public; they describe the investigations in ways that enable others to repeat the investigations.
- Scientists review and ask questions about the results of other scientists' work.

Physical Science

CONTENT STANDARD B:
As a result of the activities in grades K–4, all students should develop an understanding of

- **Properties of objects and materials**
- **Position and motion of objects**
- **Light, heat, electricity, and magnetism**

DEVELOPING STUDENT UNDERSTANDING

During their early years, children's natural curiosity leads them to explore the world by observing and manipulating common objects and materials in their environment. Children compare, describe, and sort as they begin to form explanations of the world. Developing a subject-matter knowledge base to explain and predict the world requires many experiences over a long period. Young children bring experiences, understanding, and ideas to school; teachers provide opportunities to continue children's explorations in focused settings with other children using simple tools, such as magnifiers and measuring devices.

Physical science in grades K–4 includes topics that give students a chance to increase their understanding of the characteristics of objects and materials that they encounter daily. Through the observation, manipulation, and classification of common objects, children reflect on the similarities and differences of the objects. As a result, their initial sketches and single-word descriptions lead to increasingly more detailed drawings and richer verbal descriptions. Describing, grouping, and sorting solid objects and materials is possible early in this grade range. By grade 4, distinctions between the properties of objects and materials can be understood in specific contexts, such as a set of rocks or living materials.

Young children begin their study of matter by examining and qualitatively describing objects and their behavior. The important but abstract ideas of science, such as atomic structure of matter and the conservation of energy, all begin with observing and keeping track of the way the world behaves. When carefully observed, described, and measured, the properties of objects, changes in properties over time, and the changes that occur when materials interact provide the necessary precursors to the later introduction of more abstract ideas in the upper grade levels.

Students are familiar with the change of state between water and ice, but the idea of liquids having a set of properties is more nebulous and requires more instructional effort than working with solids. Most students will have difficulty with the generalization that many substances can exist as either a liquid or a solid. K–4 students do not understand that water exists as a gas when it boils or evaporates; they are more likely to think that water disappears or goes into

Full inquiry involves asking a simple question, completing an investigation, answering the question, and presenting the results to others.

the sky. Despite that limitation, students can conduct simple investigations with heating and evaporation that develop inquiry skills and familiarize them with the phenomena.

When students describe and manipulate objects by pushing, pulling, throwing, dropping, and rolling, they also begin to focus on the position and movement of objects: describing location as up, down, in front, or behind, and discovering the various kinds of motion and forces required to control it. By experimenting with light, heat, electricity, magnetism, and sound, students begin to understand that phenomena can be observed, measured, and controlled in various ways. The children cannot understand a complex concept such as energy. Nonetheless, they have intuitive notions of energy—for example, energy is needed to get things done; humans get energy from food. Teachers can build on the intuitive notions of students without requiring them to memorize technical definitions.

Sounds are not intuitively associated with the characteristics of their source by younger K–4 students, but that association can be developed by investigating a variety of concrete phenomena toward the end of the K–4 level. In most children's minds, electricity begins at a source and goes to a target. This mental model can be seen in students' first attempts to light a bulb using a battery and wire by attaching one wire to a bulb. Repeated activities will help students develop an idea of a circuit late in this grade range and begin to grasp the effect of more than one battery. Children cannot distinguish between heat and temperature at this age; therefore, investigating heat necessarily must focus on changes in temperature.

As children develop facility with language, their descriptions become richer and include more detail. Initially no tools need to be used, but children eventually learn that they can add to their descriptions by measuring objects—first with measuring devices they create and then by using conventional measuring instruments, such as rulers, balances, and thermometers. By recording data and making graphs and charts, older children can search for patterns and order in their work and that of their peers. For example, they can determine the speed of an object as fast, faster, or fastest in the earliest grades. As students get older, they can represent motion on simple grids and graphs and describe speed as the distance traveled in a given unit of time.

GUIDE TO THE CONTENT STANDARD
Fundamental concepts and principles that underlie this standard include

PROPERTIES OF OBJECTS AND MATERIALS

- Objects have many observable properties, including size, weight, shape, color, temperature, and the ability to react with other substances. Those properties can be measured using tools, such as rulers, balances, and thermometers.
- Objects are made of one or more materials, such as paper, wood, and metal. Objects can be described by the properties of the materials from which they are made, and those properties can be used to separate or sort a group of objects or materials.
- Materials can exist in different states—solid, liquid, and gas. Some common materials, such as water, can be changed from one state to another by heating or cooling.

POSITION AND MOTION OF OBJECTS

- The position of an object can be described by locating it relative to another object or the background.
- An object's motion can be described by tracing and measuring its position over time.
- The position and motion of objects can be changed by pushing or pulling. The size of the change is related to the strength of the push or pull.
- Sound is produced by vibrating objects. The pitch of the sound can be varied by changing the rate of vibration.

LIGHT, HEAT, ELECTRICITY, AND MAGNETISM

- Light travels in a straight line until it strikes an object. Light can be reflected by a mirror, refracted by a lens, or absorbed by the object.
- Heat can be produced in many ways, such as burning, rubbing, or mixing one substance with another. Heat can move from one object to another by conduction.
- Electricity in circuits can produce light, heat, sound, and magnetic effects. Electrical circuits require a complete loop through which an electrical current can pass.
- Magnets attract and repel each other and certain kinds of other materials.

Reprinted with permission from the *National Science Education Standards*, National Research Council, Washington, D.C.: National Academy Press; 1996.

Content Standards: 5–8

Science as Inquiry

CONTENT STANDARD A:
As a result of activities in grades 5–8, all students should develop
- **Abilities necessary to do scientific inquiry**
- **Understandings about scientific inquiry**

DEVELOPING STUDENT ABILITIES AND UNDERSTANDING

Students in grades 5–8 should be provided opportunities to engage in full and in partial inquiries. In a full inquiry students begin with a question, design an investigation, gather evidence, formulate an answer to the original question, and communicate the investigative process and results. In partial inquiries, they develop abilities and understandings of selected aspects of the inquiry process. Students might, for instance, describe how they would design an investigation, develop explanations based on scientific information and evidence provided through a classroom activity, or recognize and analyze several alternative explanations for a natural phenomenon presented in a teacher-led demonstration.

Students in grades 5–8 can begin to recognize the relationship between explanation and evidence. They can understand that background knowledge and theories guide the design of investigations, the types of observations made, and the interpretations of data. In turn, the experiments and investigations students conduct become experiences that shape and modify their background knowledge.

With an appropriate curriculum and adequate instruction, middle-school students can develop the skills of investigation and the understanding that scientific inquiry is guided by knowledge, observations, ideas, and questions. Middle-school students might have trouble identifying variables and controlling more than one variable in an experiment. Students also might have difficulties understanding the influence of different variables in an experiment—for example, variables that have no effect, marginal effect, or opposite effects on an outcome.

Teachers of science for middle-school students should note that students tend to center on evidence that confirms their current beliefs and concepts (i.e., personal explanations), and ignore or fail to perceive evidence that does not agree with their current concepts. It is important for teachers of science to challenge current beliefs and concepts and provide scientific explanations as alternatives.

Several factors of this standard should be highlighted. The instructional activities of a scientific inquiry should engage students in identifying and shaping an understanding of the question under inquiry. Students should know what the question is asking, what background knowledge is being used to frame the question, and what they will have to do to answer the question. The students' questions should be relevant and meaningful for them. To help focus investigations, students should frame questions, such as "What do we want to find out about...?", "How can we make the most accurate observations?", "Is this the best way to answer our questions?", and "If we do this, then what do we expect will happen?"

Students in grades 5–8 can begin to recognize the relationship between explanation and evidence.

The instructional activities of a scientific inquiry should involve students in establishing and refining the methods, materials, and data they will collect. As students conduct investigations and make observations, they should consider questions such as "What data will answer the question?" and "What are the best observations or measurements to make?" Students should be encouraged to repeat data-collection procedures and to share data among groups.

In middle schools, students produce oral or written reports that present the results of their inquiries. Such reports and discussions should be a frequent occurrence in science programs. Students' discussions should center on questions, such as "How should we organize the data to present the clearest answer to our question?", "How should we organize the evidence to present the clearest answer to our question?", or "How should we organize the evidence to present the strongest explanation?" Out of the discussions about the range of ideas, the background knowledge claims, and the data, the opportunity arises for learners to shape their experiences about the practice of science and the rules of scientific thinking and knowing.

The language and practices evident in the classroom are an important element of doing inquiries. Students need opportunities to present their abilities and understanding and to use the knowledge and language of science to communicate scientific explanations and ideas. Writing, labeling drawings, completing concept maps, developing spreadsheets, and designing computer graphics should be a part of the science education. These should be presented in a way that allows students to receive constructive feedback on the quality of thought and expression and the accuracy of scientific explanations.

This standard should not be interpreted as advocating a "scientific method." The conceptual and procedural abilities suggest a logical progression, but they do not imply a rigid approach to scientific inquiry. On the contrary, they imply codevelopment of the skills of students in acquiring science knowledge, in using high-level reasoning, in applying their existing understanding of scientific ideas, and in communicating scientific information. This standard cannot be met by having the students memorize the abilities and understandings. It can be met only when students frequently engage in active inquiries.

GUIDE TO THE CONTENT STANDARD

Fundamental abilities and concepts that underlie this standard include

ABILITIES NECESSARY TO DO SCIENTIFIC INQUIRY

IDENTIFY QUESTIONS THAT CAN BE ANSWERED THROUGH SCIENTIFIC INVESTIGATIONS. Students should develop the ability to refine and refocus broad and ill-defined questions. An important aspect of this ability consists of students' ability to clarify questions and inquiries and direct them toward objects and phenomena that can be described, explained, or predicted by scientific investigations. Students should develop the ability to identify their questions with scientific ideas, concepts, and quantitative relationships that guide investigation.

DESIGN AND CONDUCT A SCIENTIFIC INVESTIGATION. Students should develop general abilities, such as systematic observation, making accurate measurements, and identifying and controlling variables. They should also develop the ability to clarify their ideas that are influencing and guiding the inquiry, and to understand how those ideas compare with current scientific knowledge. Students can learn to formulate questions, design investigations, execute investigations, interpret data, use evidence to generate explanations, propose alternative explanations, and critique explanations and procedures.

USE APPROPRIATE TOOLS AND TECHNIQUES TO GATHER, ANALYZE, AND INTERPRET DATA. The use of tools and techniques, including mathematics, will be guided by the question asked and the investigations students design. The use of computers for the collection, summary, and display of evidence is part of this standard. Students should be able to access, gather, store, retrieve, and organize data, using hardware and software designed for these purposes.

DEVELOP DESCRIPTIONS, EXPLANATIONS, PREDICTIONS, AND MODELS USING EVIDENCE. Students should base their explanation on what they observed, and as they develop cognitive skills, they should be able to differentiate explanation from description—providing causes for effects and establishing relationships based on evidence and logical argument. This standard requires a subject matter knowledge base so the students can effectively conduct investigations, because developing explanations establishes connections between the content of science and the contexts within which students develop new knowledge.

THINK CRITICALLY AND LOGICALLY TO MAKE THE RELATIONSHIPS BETWEEN EVIDENCE AND EXPLANATIONS. Thinking critically about evidence includes deciding what evidence should be used and accounting for anomalous data. Specifically, students should be able to review data from a simple experiment, summarize the data, and form a logical argument about the cause-and-effect relationships in the experiment. Students should begin to state some explanations in terms of the relationship between two or more variables.

RECOGNIZE AND ANALYZE ALTERNATIVE EXPLANATIONS AND PREDICTIONS. Students should develop the ability to listen to and respect the explanations proposed by other students. They should remain open to and acknowledge different ideas and explanations, be able to accept the skepticism of others, and consider alternative explanations.

COMMUNICATE SCIENTIFIC PROCEDURES AND EXPLANATIONS. With practice, students should become competent at communicating experimental methods, following instructions, describing observations, summarizing the results of other groups, and telling other students about investigations and explanations.

USE MATHEMATICS IN ALL ASPECTS OF SCIENTIFIC INQUIRY. Mathematics is essential to asking and answering questions about the natural world. Mathematics can be used to ask questions; to gather, organize, and present data; and to structure convincing explanations.

UNDERSTANDINGS ABOUT SCIENTIFIC INQUIRY

- Different kinds of questions suggest different kinds of scientific investigations. Some investigations involve observing and describing objects, organisms, or events; some involve collecting specimens; some involve experiments; some involve discovery of new objects and phenomena; and some involve making models.
- Current scientific knowledge and understanding guide scientific investigations. Different scientific domains employ different methods, core theories, and standards to advance scientific knowledge and understanding.
- Mathematics is important in all aspects of scientific inquiry.
- Technology used to gather data enhances accuracy and allows scientists to analyze and quantify results of investigations.
- Scientific explanations emphasize evidence, have logically consistent arguments, and use scientific principles, models, and theories. The scientific community accepts and uses such explanations until displaced by better scientific ones. When such displacement occurs, science advances.

- Science advances through legitimate skepticism. Asking questions and querying other scientists' explanations is part of scientific inquiry. Scientists evaluate the explanations proposed by other scientists by examining evidence, comparing evidence, identifying faulty reasoning, pointing out statements that go beyond the evidence, and suggesting alternative explanations for the same observations.
- Scientific investigations sometimes result in new ideas and phenomena for study, generate new methods or procedures for an investigation, or develop new technologies to improve the collection of data. All of these results can lead to new investigations.

Physical Science

CONTENT STANDARD B:
As a result of their activities in grades 5–8, all students should develop an understanding of

- **Properties and changes of properties in matter**
- **Motions and forces**
- **Transfer of energy**

DEVELOPING STUDENT UNDERSTANDING

In grades 5–8, the focus on student understanding shifts from properties of objects and materials to the characteristic properties of the substances from which the materials are made. In the K–4 years, students learned that objects and materials can be sorted and ordered in terms of their properties. During that process, they learned that some properties, such as size, weight, and shape, can be assigned only to the object while other properties, such as color, texture, and hardness, describe the materials from which objects are made. In grades 5–8, students observe and measure characteristic properties, such as boiling points, melting points, solubility, and simple chemical changes of pure substances and use those properties to distinguish and separate one substance from another.

Students usually bring some vocabulary and primitive notions of atomicity to the science class but often lack understanding of the evidence and the logical arguments that support the particulate model of matter. Their early ideas are that the particles have the same properties as the parent material; that is, they are a tiny piece of the substance. It can be tempting to introduce atoms and molecules or improve students' understanding of them so that particles can be used as an explanation for the properties of elements and compounds. However, use of such terminology is premature for these students and can distract from the understanding that can be gained from focusing on the observation and description of macroscopic features of substances and of physical and chemical reactions. At this level, elements and compounds can be defined operationally from their chemical characteristics, but few students can comprehend the idea of atomic and molecular particles.

In grades 5–8, students observe and measure characteristic properties, such as boiling and melting points, solubility, and simple chemical changes of pure substance, and use those properties to distinguish and separate one substance from another.

The study of motions and the forces causing motion provide concrete experiences on which a more comprehensive understanding of force can be based in grades 9–12. By using simple objects, such as rolling balls and mechanical toys, students can move from qualitative to quantitative descriptions of moving objects and begin to describe the forces acting on the objects. Students' everyday experience is that friction causes all moving objects to slow down and stop. Through experiences in which friction is reduced, students can begin to see that a moving object with no friction would continue to move indefinitely, but most students believe that the force is still acting if the object is moving or that it is "used up" if the motion stops. Students also think that friction, not inertia, is the principal reason objects remain at rest or require a force to move. Students in grades 5–8 associate force with motion and have difficulty understanding balanced forces in equilibrium, especially if the force is associated with static, inanimate objects, such as a book resting on the desk.

The understanding of energy in grades 5–8 will build on the K–4 experiences with light, heat, sound, electricity, magnetism, and the motion of objects. In 5–8, students begin to see the connections among those phenomena and to become familiar with the idea that energy is an important property of substances and that most change involves energy transfer. Students might have some of the same views of energy as they do of force—that it is associated with animate objects and is linked to motion. In addition, students view energy as a fuel or something that is stored, ready to use, and gets used up. The intent at this level is for students to improve their understanding of energy by experiencing many kinds of energy transfer.

GUIDE TO THE CONTENT STANDARD

Fundamental concepts and principles that underlie this standard include

PROPERTIES AND CHANGES OF PROPERTIES IN MATTER

- A substance has characteristic properties such as density, a boiling point, and solubility, all of which are independent of the amount of the sample. A mixture of substances often can be separated into the original substances using one or more of the characteristic properties.
- Substances react chemically in characteristic ways with other substances to form new substances (compounds) with different characteristic properties. In chemical reactions, the total mass is conserved. Substances often are placed in categories or groups if they react in similar ways; metals is an example of such a group.
- Chemical elements do not break down during normal laboratory reactions involving such treatments as heating, exposure to electric current, or reaction with acids. There are more than 100 known elements that combine in a multitude of ways to produce compounds, which account for the living and nonliving substances that we encounter.

MOTIONS AND FORCES

- The motion of an object can be described by its position, direction of motion, and speed. That motion can be measured and represented on a graph.
- An object that is not being subjected to a force will continue to move at a constant speed and in a straight line.
- If more than one force acts on an object along a straight line, then the forces will reinforce or cancel one another, depending on their direction and magnitude. Unbalanced forces will cause changes in the speed or direction of an object's motion.

TRANSFER OF ENERGY

- Energy is a property of many substances and is associated with heat, light, electricity, mechanical motion, sound, nuclei, and the nature of a chemical. Energy is transferred in many ways.
- Heat moves in predictable ways, flowing from warmer objects to cooler ones, until both reach the same temperature.
- Light interacts with matter by transmission (including refraction), absorption, or scattering (including reflection). To see an object, light from that object—emitted or scattered from it—must enter the eye.
- Electrical circuits provide a means of transferring electrical energy when heat, light, sound, and chemical changes are produced.
- In most chemical and nuclear reactions, energy is transferred into or out of a system. Heat, light, mechanical motion, or electricity might all be involved in such transfers.
- The sun is a major source of energy for changes on the earth's surface. The sun loses energy by emitting light. A tiny fraction of that light reaches the earth, transferring energy from the sun to the earth. The sun's energy arrives as light with a range of wavelengths, consisting of visible light, infrared, and ultraviolet radiation.

Reprinted with permission from *Inquiry and the National Science Education Standards: A Guide for Teaching and Learning*, National Research Council, Washington, D.C.: National Academy Press; 1996.

Frequently Asked Questions About Inquiry

Science teachers, administrators, and teacher educators (both preservice and inservice) often face difficult questions about inquiry-based teaching and learning. Many of these questions they raise themselves. Others come from teachers, administrators, preservice teachers, students, and parents who are unfamiliar with this perspective on learning and teaching science. This chapter presents answers to some of the most commonly asked questions.

Q *In inquiry-based teaching, is it ever okay to tell students the answers to their questions?*

A Yes. Understanding requires knowledge, and not all the knowledge that is needed can be acquired by inquiry. Decisions about how to respond to students' questions depend on the teacher's goals and the context of the discussion. For example, a student may pose the question "What is the boiling point of water at sea level?" One way to respond to that question would be to set up a simple investigation to find out. The investigation could set the stage for more complex inquiries. If learning to use reference material is important, a teacher might have the student look up the information. Or, if there is a higher priority for how the student spends his or her time, the teacher could simply provide the answer.

The important point is that investigations lead to deeper understanding and greater transfer of knowledge. Decisions about responding to students' questions should reflect that fact.

Q *Should a teacher ever say "no" to an investigation that students propose themselves?*

A Yes. As noted in the previous answer, a teacher's response should depend on his or her goals for the students. What might they learn if they conducted the inquiry? Are there cost or safety concerns that might weigh against doing a particular investigation? What topics and approaches are most feasible in light of the school science curriculum and guiding standards? Would it be best for students to design their own investigations or conduct investigations proposed either by the teacher or provided by the instructional materials?

A large number of learning outcomes, particularly inquiry abilities, are best learned through investigations, and those motivated by students' own questions can be invaluable learning opportunities. Students also learn the characteristics of questions that can be properly investigated if they have opportunities to pose and investigate questions. One approach might be for teachers to ask students (or help them determine) what learning goals they will achieve by pursuing their questions and which goals they will not achieve.

The fact that students are motivated to ask questions and inquire into them is an indication that the teacher is making science relevant and exciting. But not all investigations that students propose will be worth pursuing.

Is it more important for students to learn the abilities of scientific inquiry or the scientific concepts and principles?

They need to learn both. Furthermore, as the *National Science Education Standards* make clear, these are equally important learning outcomes that support each other.

In many teaching and learning sequences, students employ inquiry abilities to develop understanding of scientific concepts. Sometimes teachers assume that students develop inquiry abilities just because they use them. However, there is no guarantee of this. Instead, teachers have to work to ensure a proper balance between learning scientific concepts and inquiry abilities.

The development of inquiry abilities should be an explicit student learning outcome. Teachers can select specific abilities on which to focus and develop strategies to achieve those outcomes.

Learning science content and improving inquiry abilities can be symbiotic. Scientific concepts and inquiry abilities switch from primary to secondary focus and back again as needed to promote the effective integration of both. Also, research describes expertise as knowing both the subject matter content (the "big ideas" of the disciplines) and the ways of inquiring into new questions—and it makes the case for teaching both.

How can students do a science investigation before they have learned the vocabulary words with which to describe the results?

Scientific investigations, whether conducted by students or scientists, begin with observations of something interesting or perplexing, which lead to scientific questions, and then to reflections on what the person already knows about the question. It may seem that students need some concepts and vocabulary to begin, but investigations can be designed and carried out without knowing all the specific terms and definitions involved. In fact, the observations, data collection, and analysis involved in an investigation generally provide the context for developing operational definitions, science concepts, inquiry abilities, and an understanding of scientific inquiry, which can later be associated with names or "vocabulary."

Knowing vocabulary does not necessarily help students develop or understand explanations. Rather, once students begin to build and understand explanations for their observations, the proper names and definitions associated with those events become useful and meaningful. In essence, words become symbols for their understanding of the phenomena. As a result, definitions based on direct experience more often result in understanding than just memorizing words.

The issue of vocabulary development is particularly relevant to working with students who are English-language learners. Teachers of these students need to pay special attention to whether assessment of students' science knowledge is confounded by their use of the language, and to how student learning is supported when their language skills are just developing. As noted in research synthesized by Fradd and Lee (1999), when formulating their teaching strategies, teachers need to consider how students of diverse cultures and languages think about science, the experiences they have had in learning science, and, ultimately, how to structure new science learning experiences to optimize students' opportunities to learn important science concepts and inquiry abilities. The degree of structure given to lessons and the amount of direct "teaching" of inquiry skills need to depend on teachers' keen assessment of students' language development, current science knowledge, skills, beliefs, and cultural orientations (Fradd and Lee, 1999).

Why did the Standards choose to leave out the science process skills such as observing, classifying, predicting, and hypothesizing?

The "process skills" emphasized in earlier science education reforms may appear to be missing from the *Standards*, but they are not. Rather, they are integrated into the broader abilities of scientific inquiry. As the *Standards* point out, "The standards on inquiry highlight the abilities of inquiry and the development of an understanding about scientific inquiry. Students at all grade levels and in every domain of science should have the opportunity to use scientific inquiry and develop the ability to think and act in ways associated with inquiry, including asking questions, planning and conducting investigations using appropriate tools and techniques to gather data, thinking critically and logically about relationships between evidence and explanations, constructing and analyzing alternative explanations, and communicating scientific arguments" (National Research Council, 1996, p. 105). The *Standards* thus include the "processes of science" and require that students combine those processes and scientific knowledge to develop their understanding of science.

Q *Do the Standards imply that teachers should use inquiry in every lesson?*

A

No. In fact, the *Standards* emphasize that many teaching approaches can serve the goal of learning science: "Although the *Standards* emphasize inquiry, this should not be interpreted as recommending a single approach to science teaching. Teachers should use different strategies to develop the knowledge, understandings, and abilities described in the content standards. Conducting hands-on science activities does not guarantee inquiry, nor is reading about science incompatible with inquiry" (National Research Council, 1996, p. 23).

Everyone knows that investigations often take longer than other ways of learning, and there are simply not enough hours or days in the school year to learn everything through inquiry. The challenge to the teacher is to make the most judicious choices about which learning goals can be best reached through inquiry (remembering that deep understanding is most likely to result from inquiry), and what the nature of that inquiry should be. Other teaching strategies can come into play for other learning goals.

Q *How can teachers cover everything in the curriculum if they use inquiry-oriented materials and teaching methods?*

A

As noted in the previous question, the *Standards* do not suggest that all science should be learned through inquiry. However, investigations are important ways to promote deep understanding of science content and the only way to help students practice inquiry abilities. So there is still the issue of coverage vs. learning strategy to address.

Analysis of data collected in the Third International Mathematics and Science Study (TIMSS) reveals that the typical U.S. eighth-grade science textbook includes about 65 topics. A similarly large number of science topics appears yearly in state and local science standards and curriculum guides. Teachers, understandably, feel obligated to teach all of the topics called for in their local science curriculum. The result can be the "mile wide and inch deep" curriculum often decried in U.S. education. Furthermore, research shows that this "cover everything" approach provides few opportunities for students to acquire anything but surface knowledge on any topic (Schmidt et al., 1997).

There are several steps that teachers and administrators can take to deal with this problem. They can renegotiate the expectations embodied in the curriculum. They can carefully select a few areas to emphasize, spending more time teaching those areas through inquiry. They can carefully analyze the curriculum expectations and combine several learning outcomes in lessons and units. They can work with other grade-level teachers to eliminate the redundancies that often exist in a curriculum, but rarely deepen understanding. If they teach subjects other than science, they can integrate science outcomes into other subject areas (for example, presenting the findings of an investigation in a language arts lesson).

Teachers and administrators can be helped by district and state decision-makers who can reduce the number of topics that teachers are required to teach.

Q

How much structure and how much freedom should teachers provide in inquiry-oriented science lessons?

A

The type and amount of structure can vary depending on what is needed to keep students productively engaged in pursuit of a learning outcome. Students with little experience in conducting scientific inquiries will probably require more structure. For example, a teacher might want to select the question driving an investigation. She or he also might decide to provide a series of steps and procedures for the students guided by specific questions and group discussion. The instructional materials themselves often provide questions, suggestions, procedures, and data tables to guide student inquiry.

As students mature and gain experience with inquiry, they will become adept at clarifying good questions, designing investigations to test ideas, interpreting data, and forming explanations based on data. With such students, the teacher still should monitor by observation, ask questions for clarification, and make suggestions when needed. Often teachers begin the school year providing considerable structure and then gradually provide more opportunities for student-centered investigations.

Many teachers in the primary grades have considerable success with whole-class projects. An example is a class experiment to answer the question: "What is the 'black stuff' on the bottom of the aquarium?" Guided by the teacher, the students can focus and clarify the question. They can ponder where the "black stuff" came from based on their prior knowledge of goldfish, snails, and plants. Using their prior knowledge, the students then can propose explanations and decide what they need to set up a fair test. How many aquariums will they need? What will be in each aquarium? What are they looking for? How will they know when they have answered the question? After a number of well-structured whole-class inquiries with ample time to discuss procedures and process as well as conclusions and explanations, students are more prepared to design and conduct their own inquiries.

Q

How can teachers use inquiry and maintain control of their students?

A

To have productive experiences, inquiry requires considerable planning and organization on the part of both teachers and students. Teachers need to create systems for organization and management of materials and guidelines for student use of materials and conversation. Students need to learn how to work with materials in an organized fashion, communicate their ideas with one another, listen to each other's ideas with respect, and accept responsibility for their own learning. In addition, it always is helpful when students know what is expected of them in terms of behavior and performance. As students become collaborators, they recognize the conditions for progress themselves and need less external control.

Q

How much do teachers need to know about inquiry and about science subject matter to teach science through inquiry?

A

The more teachers know about inquiry and about science subject matter, and the more they themselves are effective inquirers, the better equipped they are to engage their students in inquiries that will help them understand scientific concepts and inquiry. It generally does not work for teachers to stay one step ahead of the students when using an inquiry-oriented program.

However, to a certain extent, teachers can develop their own understanding through inquiry as they investigate with their students and participate in professional development programs. Teachers also can consult with other teachers to learn more about a topic, refer to science background material printed in teachers guides, participate in professional development, and invite into the classroom parents, scientists, and others who have expertise to help in learning about the topic. Like their students, teachers should view themselves as learners, being eager to try new ways of teaching and extend and sharpen their subject matter knowledge. And they should use their own teaching to inquire about how to improve it, so that their ability to teach through inquiry increases in each successive year.

What can teachers do who are provided only traditional instructional materials?

Teachers who want their students to learn to inquire and to learn through inquiry are hampered if their materials are text-based and focus students on memorizing scientific laws and terminology. However, a teacher's curriculum is not defined by the materials alone, but more broadly by what students focus their attention on, how they learn, and how and on what they are assessed. Teachers can use the *Standards* to determine goals for their students and decide which pieces of their materials they can use to help students reach those goals. They can consider decreasing the "cookbook" nature of whatever "labs" or hands-on activities are included with their materials and resequencing them to come before the readings or lectures so students can explore in a concrete way *before* learning the concepts and terms. Teachers can emphasize learning the major concepts and downplay the vocabulary. They can reconstruct test items to assess major science concepts, inquiry abilities, and understandings about inquiry; they can create one full and open inquiry for students to conduct for several weeks of class. And they can supplement the materials they are given with other materials they receive in professional development or from colleagues, or locate on the Web. The important thing is to determine a set of learning goals for students that reflect the *Standards* and let those guide how and what students learn.

Where can teachers get the equipment, materials, and supplies they need to teach through inquiry?

The National Science Foundation (NSF) has supported the development and field testing of a number of inquiry-oriented science curriculum programs. These science programs, complete with student and teacher guides and materials for student activities or laboratories, are now available through commercial publishers. [See *Selecting Instructional Materials: A Guide for K–12 Science* (NRC, 1999b).] Many districts that have adopted these programs operate a centralized district materials center and loan the materials to teachers. Some districts supply a certain number of kits per grade level that are housed at school sites, with consumable supplies being replenished as needed by the district.

Where districts have not adopted such programs, individual teachers and schools have developed a variety of mechanisms to provide needed materials and supplies. Some teachers develop a list of common household materials and supplies and have students collect them from home and bring them to school. Often a group of teachers at a school will collaborate on a project so they can share materials.

If inquiry is to be the norm rather than an exception, schools must realize that materials are an essential element of teaching and should devote adequate resources and organizational structures to purchase and support use of appropriate materials. Teachers should not be expected to supply the essential supplies of teaching.

Where can teacher educators obtain inquiry-oriented programs to use in preparing teachers?

Many teacher educators use curriculum materials developed for use in K–12 classrooms to help prospective students experience and learn to use inquiry-based materials. In addition, there are materials that can be used by teacher educators, at both the preservice and inservice levels, that are designed to use for teacher learning.

What barriers are encountered when implementing inquiry-oriented approaches?

In addition to the external barriers teachers face, their beliefs and values about students, teaching, and the purposes of education can impose obstacles to inquiry-oriented approaches. Research demonstrates many of the predicaments that teachers face when considering new approaches. In a cross-site analysis of schools that had successfully initiated new approaches to science and mathematics instruction, three kinds of problems were noted: technical, political, and cultural (Anderson, 1996). Technical problems included limited teaching abilities, prior commitments (for example, to a textbook), the challenges of assessment, difficulties of group work, the challenges of new teacher roles, the challenges of new student roles, and inadequate inservice education. Political problems included limited inservice education (i.e., not sustained for a sufficient number of years), parental resistance, resistance from principals and superintendents, unresolved conflicts among teachers, lack of resources, and differing judgments about justice and fairness. Cultural problems—possibly the most important because beliefs and values are central to them—included the textbook issue, views of assessment, and the "preparation ethic" (i.e., an overriding commitment to "coverage" because of a perceived need to prepare students for the next level of schooling). In addition to this study's findings, barriers experienced currently include the widespread attitude that science is not a "basic" and the lack of appropriate instructional materials, both print and hands-on.

Q *How can teachers improve their use of inquiry in science teaching?*

A Research indicates that teachers have a fairly pragmatic approach to teaching. They tend to focus on what works to involve students or manage their classrooms, rather than on melding theory and practice (Blumenfeld, 1994). Teachers anchor their understanding in classroom events and base their actions on stories and narratives more than on theories and propositional knowledge (Krajcik et al., 1994). Thus, theory, beliefs, values, and understandings are important as teachers acquire an inquiry approach, but teachers should not be expected to address such mental constructs in isolation from their teaching context.

Collaboration can be an important catalyst of change. New understandings develop and new classroom practices emerge when teachers collaborate with peers and experts. Collaboration addresses not only the technical problems of reform but cultural issues as well. As Anderson (1996) says, "Collaborative working relationships among teachers provide a very important context for the re-assessment of educational values and beliefs. In this context—where the focus is the actual work of each teacher's own students—one's values and beliefs are encountered at every turn. It is a powerful influence. Crucial reform work takes place in this context." Collaboration stimulates the reflection that is fundamental to changing beliefs, values, and understandings.

The appropriate professional development is a powerful way for teachers to improve their use of inquiry, as long as it is viewed as support for ongoing learning that is apt to take many years to change teaching practice significantly. Teachers can become wise consumers of professional development as they broaden their images and sources of learning.